THE ROUGH GUIDE to

Android™ Phones & Tablets

by
Andrew Clare

ROUGH GUIDES

www.roughguides.com

Credits

The Rough Guide to Android Phones & Tablets

Text and design: Andrew Clare
Editing: Kate Berens
Proofreading: Susanne Hillen
Production: Rebecca Short

Rough Guides Reference

Director: Andrew Lockett
Editors: Kate Berens,
Tom Cabot, Tracy Hopkins,
Matt Milton, Joe Staines

Publishing information

This second edition published May 2012 by
Rough Guides Ltd, 80 Strand, London, WC2R 0RL

Penguin Group (USA), 375 Hudson Street, NY 10014, USA
Penguin Group (India), 11 Community Centre, Panchsheel Park, New Delhi 110017, India
Penguin Group (Australia), 250 Camberwell Road, Camberwell, Victoria 3124, Australia
Penguin Group (New Zealand), 67 Apollo Drive, Rosedale, Auckland 0632, New Zealand

Rough Guides is represented in Canada by Tourmaline Editions Inc.,
662 King Street West, Suite 304, Toronto, Ontario M5V 1M7

Printed and bound in Singapore by Toppan Security Printing Pte. Ltd.
Typeset in Minion and Myriad
Cover design: Andrew Clare

Contents

contents

Security & privacy 231

Glossary 247

Index 261

Introduction

If you've picked up this book, you probably already own a phone or tablet running Android™, or you're at least thinking about getting your hands on one. You'll be in good company: since our first edition Android has become increasingly popular, appearing on more and more devices, made by an increasing number of manufacturers affiliated with the Open Handset Alliance. It's quickly become the predominant mobile phone operating system in the English-speaking world, rising up to challenge long-established mobile operating systems thanks to its openness, versatility and speed. Ice Cream Sandwich (Android 4.0) finally brings the myriad variations that resulted from its open-source nature into a more consistent, ever more cutting-edge system. Versatility is still very much on the menu though, and with this guide you'll find it easy to set up your Android-powered device exactly the way you like.

In the past, manufacturers have embraced the freedom offered by Android's open-source code to customize how their handsets look and feel, so that exactly what you'd see on screen and how buttons functioned would vary wildly between devices. As Google™ roll out Ice Cream Sandwich onto both phones and tablets, this diverse range of products seem to be converging toward a common feature set, but with no small portion of the market still running Android 2.2 and 2.3 devices, a degree of fragmentation still exists, so we've avoided attempting to provide you with anything as unwieldy as a universal manual to Android devices.

Don't worry though, as your phone or tablet's user manual already does a great job of taking you through the features and functions of your particular device, and you'll find it an invaluable resource for referring to specifics when perusing the pages of this book. If your device didn't come with a full printed copy of its manual, you may find it in PDF format on an accompanying CD, or waiting for you on the internal SD storage (see p.107), or on the manufacturer's website.

What this book does aim to do is to go beyond the proprietary nuts and bolts in your user manual and help you explore **new ways to use your Android device**, showing you some of the best apps currently kicking around on the Android Market™ and elsewhere, while helping you sift through the **technical stuff** enough to know what's going on behind that nifty touch screen. It will help you get your **email** and **social networks** connected and **synced up** with the minimum of fuss, and dishes out all kinds of **tips** from **customizing** your **home screen** to maximizing **battery power**. For those who don't own an Android phone or tablet yet but are thinking of making the leap, there's also a handy **buying guide** to help you figure out what to look for.

About this book

Text written like **this** denotes a command or label as it appears on screen. Something written like **this** refers to the name of an app or widget that can be downloaded from the Android Market or other source.

This book was written using a Samsung Galaxy Nexus phone running Android 4.0 (aka Ice Cream Sandwich – you'll find the two terms used interchangeably throughout the book). Although portions of information in this book, including app reviews, will be applicable to earlier variations of the Android platform, we cannot guarantee that everything mentioned will be accurate for your specific device. If your phone or tablet is running Android 2.3 or earlier, you may find the first edition more useful.

Throughout the book you'll find a selection of app reviews: the number in green to the far right of each app's title represents its current user rating from the Android Market. App reviews also feature QR codes (see p.256) allowing you to jump directly to the app's page in the Market and install it simply by pointing your device's camera at it (requires **Barcode Scanner** or a similar app installed).

Acknowledgements

Thanks to Kate Berens and Andrew Lockett for their help and guidance while making this book.

Thanks also to Dave Clare and Tom Barnes, and to Kenzo Fong Hing at Google, for their answers to my relentless technical queries, and to Iris Balija for her support and encouragement.

primer

All about Android

…General questions

What is Android?

Android™ is an open mobile platform developed by Google™. It lives on your phone or tablet in much the same way that a computer operating system (Windows or OS X) resides on your PC or Mac, controlling all the hardware functions and providing a stable environment for other programs such as web browsers, email clients, media players and so on. You can use it to do many of the same things you can with a computer, plus a whole lot more.

The Android platform is open-source – meaning that Google has made the source code behind the technology openly available to third-party developers – and freely customizable. Companies making devices designed to run on Android often develop their own UI, or user interface, each with a distinct look and feel and their own slant on what is essentially a common set of core features. Some, such as Amazon's Kindle Fire and Sony's Z1000 media player are more or less unrecognizable as Android devices but use the platform as a solid base for their own interfaces and functions.

So, while the display and icons on your own phone or tablet may differ from the images you see in this book, don't let it throw you too much as the basic functions will be the same.

As well as managing a device's hardware and included software, the Android platform can also run applications from third-party

The ABC of Android comes in many flavours, all of them sweet. But behind the cute branding you'll find a powerful set of features.

developers. Hundreds of thousands of these "apps" are currently downloadable from the Android Market™, allowing you to do anything from updating your Facebook status to identifying constellations in the night sky. And, as you'd expect from a Google product, Android also integrates tightly with the many Google services you may already use, such as Gmail™, Google Maps™, Google Docs™, Picasa™, YouTube™ and others.

Android has come along in leaps and bounds since its launch in 2008, with exciting new features and enhancements built into each consecutive release. As well as going by the usual release

Stuck with an older version of Android?

With new Android™ releases coming out a couple of times a year it takes manufacturers a while to port the new code to their various devices and user interfaces. If you have an older device you may find that it's stuck a couple of versions back and you probably shouldn't expect an update any time soon. Giving your phone "root" access and installing a custom ROM ported from a more recent release may be a way to add functionality and get a bit more life out of it (see p.125). Alternatively, you'll be able to add some of the functionality of later versions using a custom launcher, such as **ADW .Launcher** or **LauncherPro** (see p.123).

Launcher Pro adds extra features and a slight performance boost to your device.

numbers, these are code named alphabetically after different kinds of dessert. Up until recently, we had Cupcake, Donut, Eclair, Froyo (short for frozen yogurt) and Gingerbread for phones. Then came Honeycomb, a separate branch of the platform built specifically for tablets. To celebrate each new release, Google places a giant sculpture of its namesake on the lawn in front of their California offices.

One of the many cool new features in Android 4.0 Ice Cream Sandwich – unlock your device with the power of your own face.

So where does Ice Cream Sandwich fit into all this?

Ice Cream Sandwich is the newest version of the Android platform. It combines the best aspects of Gingerbread and Honeycomb with new features such as face detection, Android Beam™ (nearfield communication sharing, for more detail see p.101 or the Glossary on p.255) and the ability to operate as a USB host (allowing you to hook up keyboards, mice and other devices), to provide a platform that will run happily on both phones and tablets.

How does Android compare to a computer for web browsing and email?

Obviously, because of the size of the screen and keyboard (especially on phones), the physical experience of using these features initially seems quite different from using a full-size monitor with mouse and keyboard. Once you get used to browsing with a touch screen (see the gesture guide on p.48), and with Android's support for HTML5 (and currently Flash, see p.160), you'll find very few obstacles between you and your life online. Sending emails from an Android phone feels more like sending a text message, to the point where you may have to double-check whether you're doing one or the other. If you just aren't comfortable using a "soft" keyboard to type on there are handsets available which incorporate BlackBerry-style mini-keyboards. You can also expand your set-up with a Bluetooth keyboard and mouse, or with Android's new USB functionality you can even connect a standard USB keyboard and mouse to your phone or tablet if the mood takes you. Gamers will be pleased to know that it should be possible to connect most USB or HID compatible game controllers to your tablet, including those for Xbox 360 and PlayStation 3, although depending on the input offered by a particular device, you may need an adapter.

How can I connect using my Android phone?

Android™ has built-in support for Bluetooth and Wi-Fi. It supports 802.11n, currently the fastest Wi-Fi standard, though your device's hardware won't necessarily support it. Wi-Fi standards are backwards compatible, however, so you should have no problems connecting to networks that support the older 802.11b/g standards. Unless you encounter any specific issues with a particular home network set-up, there really is no need to worry about the different Wi-Fi varieties, as they generally all coexist quite happily.

You may also have 3G and 4G web access via your SIM card and mobile data provider, useful for when you want to access the web and Wi-Fi isn't available, but depending on your contract you may have a data limit (see p.26) or be charged extra for using this service.

If your carrier allows it (and unfortunately, many don't), you can also use your Android device to supply Internet to other devices, via USB tethering, Wi-Fi hotspot mode, or the new Wi-Fi Direct mode (see p.104).

Why does the display on my device look different from the ones in this book?

In Android's early days, Google left the user interface at a fairly rudimentary level and actively encouraged handset makers to add their own customizations. This has resulted in a degree of fragmentation in terms of the way Android looks and behaves on different phones. HTC, for example, has its own Sense interface, while Sony Ericsson has Rachael, and Samsung has TouchWiz. With future releases, however, Google aims to improve the native user interface with the hope that handset manufacturers and carriers will no longer feel the need to create their own. While Ice Cream Sandwich seems to provide all the necessary bells and whistles for this to take effect, at the time of writing it's difficult to predict how manufacturers will respond to the new look, but it seems as if most of them have become rather attached to their

The same version of Android running on three handsets. Although they look very different, the basic functions are more or less the same. If you find this book talking about a feature that varies from what your own device offers, check the user manual to find out if there's a way to achieve the same thing on your model.

proprietary user interfaces, and are likely to continue with them for the time being.

So if all these devices are different, what software can I expect to find on mine?

This too will vary depending on your phone or tablet's manufacturer and which native Android applications it has combined with ones developed exclusively for its own brand. The names and interfaces of these applications may differ slightly but they perform much the same functions from one device to the next. You can expect to see messaging applications including SMS and email, a web browser, camera and camcorder utilities, a music player, calendar, Google Maps, contact lists, a news reader and some basic document readers for Microsoft Office and PDF formats. You'll probably also find built-in applications for the more ubiquitous social networking services such as Facebook and Twitter. Your home screen will have some kind of an "app tray" near the bottom of the screen that you can tap to see what you've got to start with, including the **Market** app, which you'll probably find that you use quite a lot.

You can also expect to find a fair amount of bloatware on your new device, apps and games added by the manufacturer (or phone carrier) that you may not ever want to use, but which you'll be unable to remove without rooting (see p.125). You can, however, disable these apps so that they don't appear in your app tray or load into memory (see p.116).

If you're thinking about buying a legacy device close to the launch of a new Android version, websites like techcrunch.com, techradar.com, or engadget.com (pictured) can be a valuable resource for news about which devices are likely to get the new release.

How do I know if my new device will get updates to future Android releases?

While Google are fairly good at rolling out upgrades to their affiliated Nexus™ range of phones, other manufacturers and phone carriers have mixed track records for porting the latest version of the platform to anything but their flagship models. Back in May 2011, Google made some headway in rectifying this situation, announcing a new deal they'd struck with hardware manufacturers and phone carriers guaranteeing that compatible hardware would receive timely updates for at least 18 months after their release. This looks like good news on the face of it, but obviously a device's compatibility with future software versions is something of an unknown quantity, so how this agreement plays out over the coming year remains to be seen.

Buying guide
Which model? Where from?

So you're ready to jump in and get yourself an Android™ phone or tablet. Which model should you get? What kind of phone/data contract do you need? Where should you buy it? The next few pages will help you figure out what your options are.

What should I look for?

It's always best to do some research of your own before buying a product or service, especially for something as complex and multifunctional as a tablet or smartphone. Read reviews online and check the forums at androidcommunity.com for the make and model you have in mind to see if there are any potential issues you should know about. You can also compare reviews at specialist technology sites, like the ones below:

▶ **Android and Me** androidandme.com

▶ **Engadget** engadget.com

▶ **Recombu** recombu.com

It's also worth checking Tech Radar's always up-to-date list of the twelve best Android tablets:

▶ **Tech Radar** goo.gl/U70mN

The main areas to consider when choosing a device are:

▶ **Brand:** Each of the different manufacturers – Samsung, Sony, HTC, et al. – have their own take on the Android experience (which also varies between their individual models). If you can visit a store in person, it's worth playing around with units made by all the major companies to get hands-on experience of each user interface.

▶ **Processor (CPU) speed:** Measured in gigahertz (or GHz), this is how fast your phone can process information. Dual- and quad-core processors are also now available and offer a dramatic improvement for certain applications. As an absolute minimum you need a single-core 1GHz CPU. Higher speed and multiple cores will obviously give you better performance.

▶ **RAM and SD storage:** Measured in megabytes (or MB), this is the memory that your device uses to run the Android platform and any applications you're using. More is generally better, and with apps and games beginning to creep into double-megabyte figures you can quickly find yourself running out of space. 1GB (1024 MB) should be adequate. Your device will usually also have internal SD storage for your photos, music and other media. Usually this will come in 8, 16, 32 and 64-gigabyte flavours, with higher numbers obviously meaning more space. Some devices also feature an SD expansion slot, for more storage and greater flexibility.

▶ **Future-proofing:** Which manufacturers (and to an extent, which mobile carriers) have a proven track record in rolling out Android updates to older devices? The 18-month update pledge (see p.20) should mean that this will be less of an issue in the future. Google™ usually pairs up with a manufacturer to make a Nexus™-branded flagship model to accompany each major new Android version (the first being the HTC-built Nexus One, followed by the Samsung-built Nexus S and more recently the Samsung Galaxy Nexus) and so far these models have received by far the earliest and most consistent updates. The 2010 Nexus One, for example, runs happily on Gingerbread (version 2.3x), while its almost identically featured cousin, the HTC Desire, remains stuck with Froyo (2.2x).

▶ **Screen size vs portability:** Large screens are obviously nicer to look at and better for watching video, but does your phone still fit in your pocket? Will your tablet fit in your bag? If you can, go into a shop and feel the weight of a few different models as this also varies wildly.

▶ **Screen type:** Screen quality also varies a lot; you'll want to check the resolution and viewing angle. If you don't fancy hunting around for shade so you can peer at your phone outdoors, look for high brightness and contrast ratios. Samsung's Super AMOLED screens currently lead the field in this area.

▶ **Keyboard:** Are you happy to do most of your typing with a virtual (soft) keyboard or do you need a physical keyboard that slides out from under the screen? It may be worth trying out both in a store, and figuring out whether the extra bulk of a keyboard is something you want to carry around everywhere in exchange for potentially faster typing. Some tablets, like Asus' EEE Pad Transformer, come with a detachable keyboard, making your tablet feel more like a netbook or laptop.

▶ **Camera:** How good is the device's camera? Don't just look at the number of megapixels it can capture; the quality of the sensor itself plays a major part in image quality. Camera quality varies wildly between devices so if this is important to you, a little research won't hurt and a quick Internet trawl can turn up reviews and images taken with the device you're considering. Do you want to make video calls or use the face-lock feature? You'll need a device with a front-facing camera. Some models, like LG's Optimus Pad, even have dual cameras which can shoot photos and movies in 3D.

▶ **Battery life:** This varies a lot between products and is worth looking into if you don't want to be chained to a wall socket the whole time (although there are ways to optimize it; see p.117).

▶ **Connectivity:** Does the device have an HDMI (High Definition Multimedia Interface) video output? Are the USB ports USB 2.0 or the new, faster USB 3.0? Also worth considering is whether the ports are mini versions or full-size, if easy compatibility with any of your existing USB peripherals is important to you. Adapters are cheap, though, so this isn't necessarily a deal-breaker.

▶ **Mobile antenna:** Is your tablet equipped with a 3G/4G/EDGE antenna for mobile data or does it have Wi-Fi only? If you'll need access to the Internet while out and about and don't want to rely on the availability of wireless hotspots, this may be an important consideration.

▶ **Other hardware:** Some newer devices feature gizmos such as internal barometers and NFC technology (see p.101). You'll be the best judge of which concerns are your priorities, but it's worth investigating what the options are within your price bracket.

▶ **Android version:** Obviously most recent is best but don't expect your new device to be running the latest version just because it's a recent model, as phones and tablets running much older versions of the platform are still in circulation (see Future-proofing, p.22).

Where to buy

Once you've decided on your preferred model, try a price-comparison agent such as, in the UK:

▶ **Kelkoo** kelkoo.co.uk

▶ **Shopping.com** uk.shopping.com

In the US, price-comparison agents include:

▶ **Google™ Product Search** google.com/products

▶ **PriceWatch** pricewatch.com

Some online retailers tend to make quick deliveries and offer a reliable returns service, including the best-known of all:

▶ **Amazon UK** amazon.co.uk

▶ **Amazon US** amazon.com

Mobile data for tablets

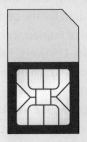 If you've decided to opt for a tablet equipped with a mobile antenna, you'll need to get a tablet broadband SIM, which is similar to a mobile phone SIM and allows access to 3G and 4G mobile data networks. You might think you could just put the SIM from your phone into your tablet to make use of any data allowance you may already have with your phone carrier, but unfortunately this isn't the case: you'll need a dedicated tablet SIM for mobile data use. Tablet SIMs are available from most of the major phone networks, usually on a pay-as-you-go per-megabyte basis, or for a rolling monthly fee. Data allowances differ depending on your chosen provider and plan, so shop around for the best deal. Different tablet models require different types of SIM, so be careful you choose the right deal for your device.

If buying a phone, you can take a contract with a network provider (in which case the cost of the phone will usually be subsidized or waived completely) or buy it outright and use a pay-as-you-go SIM from the carrier of your choice. It's worth considering how much you use your current phone and whether you really need to be tied in to an expensive 12-, 18- or 24-month contract, or if you'd like the relative freedom of pay-as-you-go. You should also investigate whether your potential network of choice has decent coverage in your area.

Signing up for a contract

When choosing a contract there are some obvious factors to consider: What's the monthly cost? How many call minutes and free texts do you get? What's the monthly data allowance? Are there any extras? Which deal you go for depends on how you're likely to use your phone: are you more of a talker or a texter? If you're a heavy Internet user, or plan to make extensive use of your phone's navigation abilities and other web services, then it's worth shopping around for the best data deal. Unlimited data deals have gone in and out of fashion with the major providers but seem to be on the up again. US users, for example, can now get a deal with T-Mobile via Walmart which offers fewer talk minutes in favour of unlimited mobile data.

The main network carriers have branches in most major towns and cities, so you can easily walk in and pick up a phone on the spot. There are also one-stop shops where you can peruse offers from various carriers in one place and pick the one that suits you best:

In the UK:

▶ **Carphone Warehouse** carphonewarehouse.com

▶ **Phones4u** phones4u.co.uk

In the US:

▶ **Walmart** walmart.com

▶ **Best Buy** bestbuy.com

▶ **Target** targetmobilestore.com

It's also well worth shopping online. Apart from the carriers' own websites, there are some phone comparison sites that resell the same package deals you'd get in-store but with substantial discounts or cashback incentives that could cut the cost of running your phone by up to a third over the life of a contract.

Authorized online agents are paid commissions by carriers to bring in new customers, offer a simplified credit application process and often give discounted phone prices.

In the UK:

▶ **The Phone Network** thephonenetwork.co.uk

▶ **Buy Mobile Phones** buymobilephones.net

▶ **One Stop Phone Shop** onestopphoneshop.co.uk

Using your phone or tablet overseas

Shortly after arriving in a new country your phone will detect local networks and offer you the choice to connect. From then on you can make and receive calls as normal, but check with your provider back home to see what their charges are. When it comes to foreign use of web, email, maps and other Internet-based features from your phone or tablet, you're best sticking to Wi-Fi hotspots where you can get online for free (or pay a reasonable charge for access). In many countries it's possible to connect via the local mobile data networks, but brace yourself for some astronomical data roaming fees when you get back home. If you can stomach the costs, turn **Data roaming** on from **Settings > Wireless & networks > Mobile networks**. You may also need to contact your phone company so that staff can activate data roaming at their end. A cheaper solution if your phone is unlocked (see p.29) is to purchase a compatible prepaid SIM when you get to your destination and then access networks at the local rate, or to grab one of the SIMs from maxroam.com or gosim.com, which are usable worldwide for a reasonable rate.

For more, check out Timedials SIM comparison generator: timedial.net/sim-cards-price-compare, **or check** moneysavingexpert.com's **overview of your data roaming options:** goo.gl/sAKmL

In the US:

▶ **Wirefly** wirefly.com

▶ **Let's Talk** letstalk.com

▶ **A1 Wireless** a1wireless.com

Leaving your contract early

If you're in the US and want to get out of your existing contract, it may be possible to sell on the remaining months of your deal through one of several online reseller programs:

▶ **Cell Trade USA** celltradeusa.com

▶ **Cell Swapper** cellswapper.com

Or, in the UK:

▶ **Cell Swapper** uk.cellswapper.com

Used/secondhand devices

Buying a secondhand Android™ phone or tablet is much like buying any other piece of used electronic equipment: on the one hand you could find a bargain but you might just as easily land yourself with an overpriced paperweight. If you buy a device that's less than a year old, it should still be within warranty, so you'd be able to get it repaired for free if it doesn't work properly – even if the item in question was purchased in a different country.

Whatever you buy, it's good to see it in action before parting with any cash, but remember that this won't tell you everything. If a device has been used a lot, for example, the battery might be on its last legs and soon need replacing, adding indirectly to the cost.

If you buy on eBay, you'll get loads of choice and a certain level of protection against getting sold a dud. Be sure to read the auction listing carefully and don't be afraid to ask the seller if you're unsure of anything. You'll often find people selling unwanted brand-new phones that they've received as part of an ongoing network contract. This can be a great way of snapping up a bargain, but be aware that the handset you're buying may well be locked to that network and will need to be unlocked if you want to use it with another.

How do I "unlock" my Android phone and use another SIM card or network?

If your phone is "locked" to a specific network and you'd like to unlock it, for example if you're travelling abroad and want to save money by using a local SIM card, or if you're at the end of your contract and want to move over to another provider on a pay-as-you-go tariff, the process will be as straightforward as unlocking any other phone.

Before you can unlock your phone you'll need to know a few bits of information:

▶ Your phone's make and model

▶ Your current network provider

▶ Your phone's IMEI code – this is your phone's number plate, usually found hiding behind its battery, or by dialling ***#06#**, or by going to **Settings > About phone > Status**.

Tip: It's good to have your device's IMEI code written down somewhere, as in the event of loss or theft you can contact your provider and have them use this number to lock your phone down. It'll also make your phone easy to identify if recovered.

Once you have this information to hand it's a simple case of browsing one of the many sites that offer unlocking codes. Visit cellunlocker.net or imeisimunlock.com, enter your information and get the unlock code to enter into your phone. A lot of sites will charge you for the service but surf around a bit and you should be able to find a free one. Check out an "unlockapedia" for more details:

▶ giffgaff.com/unlock

Insurance

It may seem a bit of a risk walking around with a flashy new Android™ device worth hundreds of pounds in your pocket, and the insurance packages offered by your network carrier may suddenly start to look like a good idea. It pays to shop around, though, as the coverage offered by a phone provider probably costs over the odds compared with what you can get from third-party insurers. In the UK, visit insurance2go.co.uk or insurance4mobiles. co.uk and explore your options. In the US, try ensquared.com or gocare.com. Policies vary in price depending on what's on offer, so consider how important the following are to you:

▶ Are you covered for calls made by someone else if your phone is stolen?

▶ Are you covered for accidental damage and loss as well as theft?

▶ Are you covered if you go abroad?

▶ Will it cover damage or loss of the device when used by other family members?

▶ Are your accessories covered?

▶ Will you need to pay an excess, and if so, how much?

Policies can also vary depending on whether you're on a pay-as-you-go or monthly contract. It's worth checking too whether your provider will send you a secondhand or refurbished replacement phone instead of a new one.

Tip: As an alternative to getting a specific insurance policy, check whether your home insurance covers your device, or whether it could be added to the policy for a small addition to your regular payment. You may also be able to upgrade your bank account to one that includes mobile phone insurance.

getting
started

Getting started

Jumping through the first few hoops

If you've owned a smartphone or tablet before, chances are you're familiar with most of the jargon used within these pages and will be able to find your own way around the device. You can probably skip ahead to the more technical stuff (p.97). If, however, you're more prone to scratching your head and throwing bewildered frowns in the general direction of new gadgets, you'll find a helpful glossary on p.249.

Switching on for the first time

First things first, open up your device and insert the battery (and SIM, if you have one) and connect the charger. If you can resist the temptation, it's a good idea to let the battery reach full charge before you switch the thing on for the first time.

The first time you switch on, you'll see a welcome screen; select your language from the drop-down menu and press **Start**. Next, you'll be prompted to sign in to your Google™ account, or to create a new one. You can skip this stage if you like, and enter your account details later by heading to **Settings > Accounts & Sync** and tapping **Add new account** at the bottom of the screen. If you don't already have a Google account and feel like you need a bit of convincing before you dive in, the box on the next page will provide all the arm-twisting you'll need.

Why do I need a Google account?

Having a Google account is going to make pretty much every aspect of your Android™ experience a lot simpler. If you're determined not to have one, it is just about possible to set up a different email account, to install your apps from non-Market services and so on, but things are going to be so much easier if you have a Google account to fall back on. Google accounts are free, easy to set up and provide a whole range of services, from Gmail™ to Calendar™, from online storage to the Android Market™, and syncing all this stuff between your phone, tablet, computer, and the cloud is a doddle.

It's well worth taking a couple of minutes to set up a Google account; it'll save you hours further on down the line, and if you do decide later on that you don't need it, you don't have to use it.

From Gmail, for example (Gmail is Google's free webmail service), you can import your contacts from wherever you keep them (Outlook, Hotmail, etc) and have them sync wirelessly among all your devices (see p.137). You can create groups, add new contacts, delete contacts and perform all other contact-related actions from either your phone's contacts list or your computer via your Gmail account, and (assuming you have an active Wi-Fi or 3G connection) one will mirror the other within a few seconds. For an in-depth look at syncing your mail and contacts via a Google account, see p.129.

Before you can sign in or create an account, you'll need some kind of Internet access: either your mobile data connection will kick in (if you have one), or you'll be switched to a Wi-Fi settings screen (see p.98).

Importing your contacts and settings

If you have your contacts stored in a Google account from a previous Android device, you'll find that these have already synced wirelessly from your Google contacts. If your previous device also had contacts synced with Facebook, Twitter or Google+™ you'll have a little extra setting up to do in order to get their photos and data resynced (see p.137), but for the most part you're done.

If you're switching phones, you'll also be able to import contacts from the old phone's SIM. Save the contacts from your old phone to your old SIM and then pop that into your new phone to copy them across. For those with a microSIM, adapters are available for around $5.

Go to the **People** app (depending on your make and model, it may be called

Android's contact manager is called People. It'll drag in pretty much all your contact data from any accounts you associate it with, so unless you love scrolling through everyone who's ever emailed or tweeted at you, spend some time assigning labels to your contacts so that you can choose which groups to display (see p.138).

Contacts or something to that effect), press the in-app **Menu** button (see p.39) and select **import/export > import from SIM card**.

Importing your apps

If this isn't your first Android phone or tablet, you'll no doubt have an arsenal of apps and games kicking around on your other device(s). In a perfect world, associating your new device with the same Google account would be enough to make all that software magically download and install itself. Well, it turns out that in this (admittedly, very minor) instance we are in fact living in a perfect world, because that's exactly what happens. More or less. You may find that your apps fail to sync automatically the first time around, especially if you had to fumble around setting up a Wi-Fi connection. Your best bet at this stage is to perform a factory reset: **Settings > Backup & reset > Factory data reset**. Alternately you can just reinstall your apps manually, from the **Market** app on your device, or from a web browser (see p.77).

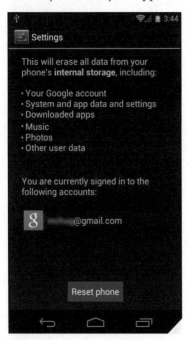

It might seem drastic, but if your device isn't behaving itself when you first start using it, a factory reset may be the best way to gently kick some sense into it.

Finding your way around

Whether you're staring blankly at a new phone or staring blankly at a new tablet, if you've just turned it on, the thing you'll most likely be staring blankly at is your home screen. Similar to a computer's desktop, it's here that you can store shortcuts, folders, widgets and apps for easy access. It has a few buttons at the bottom for the most frequently needed functions, a clock, and plenty of real estate for you to populate with all your stuff.

Unlike a computer's desktop, you and the home screen interact not via the endless shunting around of a mouse, but with multi-touch gestures (see p.48), flicking and tapping the device's touch-sensitive screen with your fingers or a stylus. The other main difference is that Android™ has multiple home screens (just how many depends on which model or launcher you're using, but it's usually at least five). You can shimmy around between these screens by swiping from left to right with your finger. On some devices you can "pinch" the home screen (see p.49) to display thumbnails of all the other screens at once. Tapping on a thumbnail takes you to that screen. If at any point you want to find your way back to your main home screen, tap the **Home** button, the vaguely house-shaped button in the centre (see below). It'll take you back home from any screen, any app or game, any time you want.

Tip: Depending on your device's manufacturer (particularly if the device in question is a phone), many of the features described in this section will look quite different from those discussed here. Don't panic though, as this is simply brand customization; the basic functions are all there and for the most part will work in the same way.

The three main navigation buttons. In some apps, such as the camera, they'll shrink down to unobtrusive grey dots, while in others (when watching videos in YouTube™, for example), they hide off-screen altogether.

Navigation buttons

The **Home** button sits at the bottom of your screen (across the bottom for phones, and over to the left for tablets) between two other handy little "soft" buttons (ie, not physical buttons, but icons on the screen), the **Back** button (to the left) and the **Recent Apps** button (to the right).

The **Back** button performs the simple task of taking you back one level to where you were before, so if you're looking at a web page, tapping it will take you back to the previous page. If you're messing around with your settings, the Back button will take you back to the previous settings window. It also acts like a quit button, taking you out of an app and back to the home screen.

Hands up who hates multitasking! Okay, now you'll have to keep your hand up forever for fear of accidentally pressing the **Recent apps** button.

The **Home** button, as discussed above, takes you home from anywhere, any time (it's a shame they haven't invented these for real life yet).

Tapping on the **Recent Apps** button (also known as the Task Switcher button) pops up a list of all the apps you've recently used. Unlike on a computer, apps don't really close when you stop using them, they just sort of hang around in the background, so you can use this button to quickly jump between them. You can flick up and down to scroll through the list, flick individual items out to the right to discard them, or long-press to bring up more options, including access to the app's info page (see p.115).

Menu button

Previous generations of Android phones usually had **Menu** buttons, which you could press for an array of options within most apps. If you're a seasoned Android user you may be wondering where they've hidden this in the new release. For the most part it's been replaced with the Action bar (see p.44) but it still appears when there are too many options to display there, acting as an overflow for less frequently accessed functions. For older apps not featuring an Action bar, the Menu button appears alongside the other soft navigation buttons, but only when there's actually a menu to be had.

It looks like this: ⋮ and usually appears in the top right of the screen on tablets, and in the bottom right on phones, although

The new **Menu** button (shown highlighted), now only appears when entirely necessary. It may also make errant appearances on some older apps but not actually do anything when tapped.

it can also pop up in various other locations within apps, such as in the message headers in Gmail™, giving you message-specific options that would otherwise eat into screen space.

Your apps & widgets are only ever a single tap away. Hit the **Market** icon (top right) to hunt for more.

App tray

The app tray contains all the apps and widgets you have installed, and is accessed by tapping the **All apps** button (shown above). Use the tabs at the top to switch between apps and widgets. You can swipe left and right to browse further. Phones usually have the app tray at the bottom of the screen, in the middle of the Docker (see below), but depending on your device it may just be a button with a little arrow on it that's built into a panel at the bottom. Tablet users will find their app tray hiding up in the top right-hand corner.

The Docker

In addition to the home screen features already listed, phone users will have a Docker (also sometimes referred to as the "docking tab" or "favourites tray") at the bottom of their screen (see above). The Docker sits just above the navigation buttons and contains the All apps button (centre), along with buttons for frequently used apps, such as the **Dialler** (far left, for when you use your phone as an actual phone), the **Browser** (far right, see p.156), **People** (second left, your contacts list, see p.136), and SMS (text) messaging (second right, p.92). As you flick between your multiple home screens, you'll notice that the Docker stays put, and appears on all of them. If you don't find the selection of apps your Docker offers especially useful you can remove them and add others instead (with the exception of the All apps button, which is non-negotiable).

Notification bar

This is a part of the home screen that provides you with all kinds of updates, from letting you know you have a missed call or new email, to informing you that you're running low on space for installing new apps.

Phone users will find the notification area at the very top of the screen. It's a black bar with a few items over to the right, usually including a clock, battery level indicator, signal meter, and so on. Other icons will come and go, letting you know Wi-Fi strength,

The Notification bar, shown here being accessed from a phone's lock screen. Tap on any of the items to head directly to the associated app. Notice the blue **Settings** icon top-centre, next to the date. Tapping that takes you straight to the main settings menu for your device. The blue × (top right) clears the decks of all your updates.

whether GPS or mobile data is in use, that kind of thing. Over to the left, notification icons will pop up now and again for missed calls, new messages, and other bits and pieces. For more information, touch on the bar and drag down. Tapping on an item will take you to the appropriate location. For example, tapping on a new message notification will open the message, tapping on a notification that an app has successfully installed will open that app, and so on. To clear notifications, use the discard gesture by swiping it off to the right. At the top of the notification area you'll also find a quick shortcut to your settings – the little "slider" icon next to the date.

On tablets, the notification area is down in the bottom right of the screen. As with phones, icons will appear to represent notifications from different apps. You can tap on these individual icons to pop up that particular notification, or bring up the whole

Android 4.0 running on a tablet with the **Notification area** (bottom right) fully open. The other key elements use the larger screen area as an opportunity to get as far away from each other as possible, with the **All apps** button hogging prime position in the top right, **Google Search™ bar** grabbing top left, and the **navigation buttons** skulking down in the bottom left corner. No sign of the highfalutin **Menu** button yet; it's probably still doing its hair.

tab by tapping on or around the clock. Individual notifications can be cleared by swiping them off to the right. Tapping on the settings icon 〓 expands the panel further to include switches for quickly changing Wi-Fi state, airplane mode, adjusting brightness, toggling auto screen rotation, and entering the settings menus.

Settings

 As mentioned, you can find the settings icon hiding in the top of the notification area, but also in your app tray (you can place it on your home screen too). Tapping on it takes you into a labyrinthine realm of menus and submenus.

Google Search™ bar

This lives at the top of the home screen, and like the Docker, stays put as you swipe between screens. At first glance it may seem like a slightly pointless web search widget that you can't get rid of, but this little search bar is deceptively useful, giving you quick access to any contact, app, web search, navigation, voice action, or just about any other item or function on your device. Tap anywhere in the bar to initiate a search, or tap the microphone icon to perform a voice command (see p.59). Once in the search window, you can define which kinds of items are searchable by tapping **Menu** > **Settings** > **Searchable items**.

Don't worry though, it's not as scary as it looks, and we'll be getting into it in more detail later on.

Action bar

The Action bar appears when you're using certain apps, materializing along the top or bottom of the display, depending on the app and screen orientation. It's populated with a handful of context-relevant buttons for the more frequently used functions or sections of an app (deleting and archiving items in Gmail,

The Action bar as it appears when reading a message in Gmail. The buttons are contextual, so depending what you're doing in any given app, you'll see the buttons most useful to you at that particular moment.

for example), occasionally including the aforementioned Menu button, which you can tap for more involved operations. If you're not sure what an icon represents, long-press it and you'll get a little tool-tip describing its use.

Lock screen

Before you get as far as the home screen to actually use any of these items, you need to navigate the lock screen. The lock screen does just as its name suggests: it locks your screen to prevent it accidentally activating while in your bag or pocket. Depending on whether or not your lock screen has been customized by the manufacturer, it may be a curved bar that you swipe down to unlock, or some icon that you need to drag over to one side. The standard lock screen in Android 4.0, however, displays a little padlock in the bottom half of the screen with a circle around it. To unlock, drag the circle over to the right, towards the unlocked padlock icon (it'll materialize as you start dragging). Alternately, to go straight to the camera if you're in a hurry to take a quick snap, drag to the left, towards the camera icon.

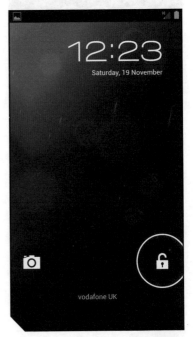

Lock screens usually display the time, date and battery status towards the top of the display.

Standard lock screen in Android 4.0. Drag the circle to the right to unlock or left to start taking snaps.

The notification bar is also accessible – just drag it down from the top as you normally would. Tapping on a notification takes you to the relevant app or service without having to unlock first.

Tip: The lock screen is basically just there to prevent your device being switched on accidentally. For a more secure option you'll need to set up a pattern lock, PIN lock or face lock (see p.238), although these will remove the straight-to-camera functionality of the default option.

WidgetLocker $2.99 (£1.87)

4.3

A crafty little app that lets you modify your lock screen with wallpapers, sliders for changing brightness and volume, and a whole host of other settings and customizations. You can place shortcuts, apps, control panels and any widgets you like for quick and easy access without having to unlock your device.

WidgetLocker puts widgets, shortcuts and custom sliders on your lock screen.

Rolling with your home screen

Now you know all about the basic functions of the home screen and buttons, what are you going to do with all that empty space? Why, fill it up of course! You can throw all manner of apps, widgets, folders and shortcuts at it, drag them around and resize them till they suit your needs.

Widgets are awesome; you can get immediate access to all kinds of live information – check your calendar and to-do list, see what music is currently playing, or scroll through tweets – just by flicking around your home screens. To harness all that widgety goodness, look for the **Widgets** tab at the top of your app tray.

Adding apps and widgets to your home screen

Adding an app to your home screen is about as simple as it could get. Tap on the app tray icon to open it up. As discussed above, you can flick left and right between screens of apps, and hit the **Widgets** tab at the top to do the same with your widgets. To add one to the home screen, simply long-press it. After about a second, the app will lift up and become highlighted. Meanwhile the app tray will fade away revealing the home screen. Drag the app (or widget) to where you'd like it to be and let go. To move it onto another screen, just drag it over to the edge of the screen and it'll shuffle over for you. Widgets often require a bit more space than apps, so if there's not enough room on your chosen screen you won't be able to place it there until you move or delete something else.

Moving and removing items

If you'd like to move an item around your home screen, simply long-press it and it'll become highlighted again. Drag to the new location and let go. You'll also

Touch screen gestures

Android™ supports a small number of touch gestures that make getting around your phone much easier:

Tap: pretty straightforward, simply tap the screen to activate any app, widget or button, to make selections from menu lists, toggle check-boxes and so on.

Double-tap: basically two short taps in rapid succession. Toggles between a zoomed-in and fit-to-screen view in many apps, including the photo gallery and web browser.

Long-press (or **long-touch**): press and hold for a second or more. Usually this brings up additional options or functions related to an object or button, for instance:

▶ Long-press a button in the Action bar to bring up a tool-tip description of what the button is for.

▶ Long-press the device's **Back** button when using the stock web browser to show your browsing history, bookmarks and saved pages.

▶ Long-press any text on a web page to select parts of it and trigger copy and paste functions (see p.55).

▶ Long-press in any text input field to switch between different keyboards if you have more than one installed, and to choose select, copy and paste functions.

▶ Long-press any item in a web browser to view options for that item (copy link, save image, bookmark link and so on).

▶ Long-press keys when using the on-screen keyboard to choose from a list of alternative characters, accents, punctuation and so on.

▶ In your calendar or any list such as your contacts or text message archive, long-press an item to show a menu of different options for that item (add event to calendar, delete contact and so on).

Pinch and **spread** (or **unpinch**): spreading your finger and thumb away from each other zooms in to pictures or text. Very useful when viewing those tiny doll's-house versions of web pages that you get when you first open them up. Pinching, naturally, does the opposite.

Drag and **swipe**: dragging means touching your finger to the screen and moving it around, for scrolling through lists and moving between your multiple home screens, the pages of an eBook or items in your email. Swipe achieves the same thing but is more of a casual flick and moves things along faster. You can also drag or swipe individual items in your notification bar or recent apps list off to the left or right to discard them.

Rotate: Most notably used in **Maps** and other navigation apps, use two fingers or finger and thumb (spaced a centimetre or more apart) to twist the view round to the desired angle.

These gesture functions can be augmented with a few apps such as Google™'s **Gesture Search** (see p.120), which allows you to draw letters on your screen to quickly home in on one of your contacts, and the **Open Gesture** and **Oftseen Gestures** apps, with which you can assign functions to gestures of your own invention.

Widget resizing – don't worry, it's got nothing to do with enhancing your manhood.

notice a **Remove** area appear at the top of the screen, drag the item up there and it'll be gone.

Resizing widgets

The neat thing about widgets in Android 4.0 is that you can now resize them, giving them as much space as they need. To resize a widget, long-press it until it lifts up ready to be moved, then, without moving it anywhere, let go. A rectangle will appear around the widget with little handles at each side. All you have to do is drag the handles out (or in) to taste.

You'll have noticed by now that there's an invisible grid that items on the home screen snap to. Widgets are no exception and will resize to this grid, ensuring there'll always be a nice tidy formation, no matter how much tweaking you do.

Adding folders

If your home screen is getting cluttered and you'd like to free up some space by grouping some items together, you can do this by making a folder. Simply long-press an item until it lifts up and drag it on top of another item. The two icons will combine to make

Arrange your apps and widgets into useful folder groups to save on home screen space. This picture shows an open folder named Google; tap on the text at the bottom to re-name it.

a folder. Tap the folder and it'll open up to reveal the items it contains. You can re-arrange items in folders by pressing and dragging them around. To remove an item from a folder, just drag it back out onto the home screen.

Making a shortcut

A shortcut is a type of widget that performs an action that you pre-set when placing it on the home screen. Shortcuts are mixed in with all your other widgets in the app tray. With the stock shortcuts you can set up one-click events for navigating back to a certain address from wherever you are, calling or messaging a frequently used contact without trawling through your contacts list, listening to a music playlist, going directly to

Staying in an unfamiliar part of town? Why not create a handy shortcut to navigate your way home (see p.222).

a particular item buried in the settings menus, and so on. You'll also find plenty of app-specific shortcuts and widgets that extend the functionality of some of your installed apps.

Changing your wallpaper

To change the wallpaper, long-press on an empty part of your home screen and select from the menu that pops up:

▶ **Live wallpapers:** Live wallpapers are animated backgrounds you can interact with. An animated waveform that responds to the sounds around you, Koi carp that swim away from your poking finger, there are plenty of pre-installed live wallpapers to choose from and hundreds more available from the Market.

▶ **Gallery:** Allows you to select a picture from your photo gallery, including any images you have stored online in accounts associated with your Google login.

▶ **Wallpapers:** Just regular old wallpapers. Select from the ones available or go fill your boots from the Market.

Tip: To automatically set wallpaper to change at regular intervals, check out the free **Wallpaper Changer** app from the Market.

Working with text

Using the on-screen keyboard

If you have a device that has a slide-out "hard" keyboard, or have a USB or Bluetooth peripheral keyboard to connect, you may prefer to use that for heavy-duty typing, but the on-screen, or "soft" keyboard, while a little frustrating at first, is surprisingly easy to get to grips with. The text prediction is very intuitive in replacing any typos resulting from the close proximity of the keys, and if you switch to horizontal mode (by simply turning your device onto its side) the keyboard will expand to fill the full width of the rotated screen and becomes easier to handle.

To start typing, just tap into a text field, web address bar, or anywhere else requiring you to enter some text. The keyboard will appear instantly and away you go.

To bring up extra characters you can press the **?123** button (bottom left). You can also long-press the appropriate key from the top row until a pop-up appears and enter numbers from there. Other extended characters such as accents can be found in the same way; if you want an á, â, ã, å, or an ä, for example, just long-press the **a** key and select it.

Text prediction will usually cover any mistakes; as you type, keep an eye on the bar in the middle of the screen, which

Extended characters in the stock keyboard. Your device may well have its own keyboard layout, with more extended characters available directly from a long-press of each key.

If you suffer from the medical condition known as "Sausage Fingers", you'll find that the on-screen keyboard gets a lot less fiddly if you flip your phone or tablet into a landscape position. For quick punctuation without waiting for a long-press, tap the full-stop (period) key and choose from the options given in the prediction bar.

throws up suggestions you can tap on to enter the entire word. The centre word is the one it'll use unless you select otherwise. Long-

press any of these suggestions for a pop-up list of even more options.

Adding new words to the dictionary

To enter a word not already in the dictionary, without being auto-corrected, type it out and select it from the suggestion list. If you tap it again it will be permanently added to your dictionary. Any unrecognized words already entered will be underlined in red within your

Long-press a word from the prediction bar for a broader range of suggestions.

entered text. You can long-press on these and choose correct alternatives or add them to the dictionary.

> **Tip**: If you have more than one keyboard installed (see custom keyboards on p.57) you can switch between them (while the keyboard is visible) from the notification bar. Just tap **Select input method**.

Selecting, copying and pasting text

To select any text, simply long-press it. The word you're pressing on will be highlighted, and you'll see a little tab either side of it (as shown here). Drag the tabs to the left or right to include more or less text in your selection.

Greetings friends!

It has been a little while since I've been in touch with you all and I wanted to give you a quick update on what Gringo has been up to. I've found email addresses for as many band members as possible, but if I have missed anyone or you know their email address has changed, please forward my email on. If you want to reply to this email, just reply to me and don't cc everyone as it will probably hack people off!

Next summer is Gringo's 15th birthday. I'm hoping to

Can I use a stylus instead of my finger?

You can, but not any old stylus. The touch screens on most Android™ devices work by recognizing the very specific electrical capacitance of your fingers, whereas most traditional stylus-touch systems tend to be pressure sensitive. There are a number of conductive "soft touch" styli (often advertised for use with iPhones) on the market and any of these should work a treat if you don't like covering your phone in greasy fingerprints or just prefer using a stylus for fiddly operations like typing on the virtual keyboard.

Once you're happy with your selection, you'll find the Action bar offers various options, usually select all, cut, copy and paste. The check-mark over to the left will just deselect everything and quit out of the copy/paste process. Some apps, such as the web browser, for example, will give you a few more options hidden behind a menu button, such as **share** and **web search**.

In addition to these options, you can also just drag your selected text around, for example if you just wanted to move a word elsewhere in the sentence.

To paste your copied text somewhere, just long-press in the area you'd like to paste into and a pointer appears under the text. You can adjust the pointer's location by dragging it. Then either hit the floating **Paste** button or tap the paste icon in the Action bar.

Selecting text shifts the focus of the Action bar to facilitate a select, copy and paste focus. The icon in the middle is the **Share** button, with a small corner triangle beneath it indicating that a list of further options sits behind the button.

SELECT ALL				

Custom keyboards

The Android Market™ hosts a shedload of custom keyboard apps with a variety of quirky functions. Some offer speedier text entry by letting you slide between letters without taking your finger off the touch screen, while others offer advanced predictive text engines that anticipate what you're going to say next with a disconcerting degree of accuracy.

Swype Free   4.5

Probably the best-known custom keyboard out there, but also the biggest pain in the posterior to install. Swype provides a novel approach to typing, whereby you just zigzag your finger around casually between the letters. Its prediction engine is surprisingly good considering the apparent sloppiness of the input method. If you need to, you can still tap out words in the regular fashion, helpful for entering place names or anything else not already in the dictionary. For an alternative, try **SlideIT**, which is $5.99 with a free 15-day trial version also available.

Swype is free but you can't get it from the Android Market. Instead, you'll need to register at beta.swype.com and follow the slightly convoluted instructions from there. Swype isn't available for all devices, but the developers frequently add new ones.

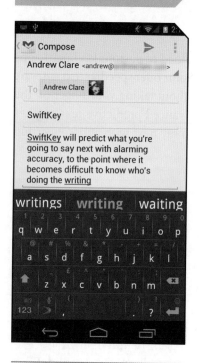

Have we known each other too long? SwiftKeyX scans your previous messages to get an idea of what you type. After a while, it seems that all you have to do is repeatedly tap the space bar to approve the suggested words in order to get your sentence out.

SwiftKeyX Free/$3.99 (£2.49)

4.5

It may look just like any other keyboard, but SwiftkeyX has an impressive text prediction engine that seems to know what you're going to type next even before you do. Instead of just auto-completing each individual word as you type, it actually learns your writing habits and attempts to predict your next word, becoming increasingly accurate the more you use it. Simply hit the space bar to approve the suggested word, or tap one of the other suggestions either side. It can optionally monitor your Gmail, Facebook and Twitter posts for greater accuracy. SwiftKeyX is available in a variety of different flavours, including a tablet-optimized version.

Using voice recognition

Dictation

Voice-to-text in Ice Cream Sandwich is more responsive than ever. To dictate, just tap into a text field as you normally would to bring up the keyboard. Now tap the microphone button and start talking. Speak more or less as you naturally would. Obviously you need to be a *little* bit clear, but there's no need to bark out each word individually or over-pronounce anything. Text should appear almost as rapidly as you speak. For punctuation, just say what you need (although it only seems to recognize "period", and not "full stop"). Any words the system is unsure of are underlined in red; long-press these to correct them with alternatives. Speech recognition relies on the Internet, so any lag you experience may be down to a poor Wi-Fi or data connection.

Hmm, that's the last time I try to dictate anything immediately after having a wisdom tooth out. Not to worry, missed words are more or less easy to correct.

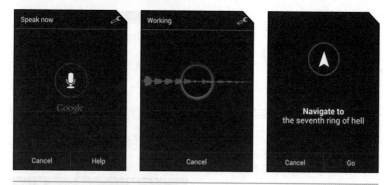

Voice actions for things like hands-free navigation are pretty neat, although you might feel a bit daft dictating a text message to someone; if you're going to speak out loud anyway, why not just phone them, say your piece and hang up?

Voice actions

Voice control in Android 4.0 is faster and more accurate than in previous releases. Simply press the microphone icon next to the search field and speak your query or command. If your device doesn't already have this facility, you can download the free **Google™ Search By Voice** app from the Market to enable voice search (and then add any custom keyboard (see p.57) that incorporates the voice button, to allow for dictation).

Speak a search query clearly towards your phone or tablet and you'll be taken to a Google search page. Speak someone's name and you'll be taken to their contact entry. You can also use commands such as "Navigate to the nearest pub", "Map of Reykjavik" and "Phone (or Call) Jim", and be taken to the appropriate map or phone function (useful if you're driving and need to make a hands-free call or get directions). You can also dictate a quick note – "Note to self: never eat anything bigger than your head"; and get some music going – "Listen to Bald Mermaid". Voice actions are developing rapidly, and can be augmented with some great apps from the Market.

She's not much to look at but Jeannie beats the other voice command apps hands down. She also, for some reason, has a man's voice in the UK version. The feature list also includes the tantalizing Voice Paint option, the example given being "Paint a forest with trees and rabbits". Repeated attempts failed to yield any results, but at least Jeannie didn't ask us to stop shouting.

Jeannie Free

4.0

Since Apple raised the bar a couple of notches with their much-touted Siri app, Android developers have been swift to respond, with a rapidly developing array of options. We tried them all, and while most just seemed to offer the same actions as Google voice search with the addition of Wolfram Alpha search capability, Jeannie was capable of so much more. As well as answering your questions with a reassuringly assertive yet strangely wobbly voice, Jeannie can change system settings, install and launch apps, make appointments, set alarms and so on. Other voice assistants worthy of your attention include **Iris**, **Edwin**, **Vlingo** and **Skyvi**.

Sharing files

Android™ makes it incredibly easy to share photos, videos, songs, map references, contacts, web links, apps, or anything else you can access with your device. You can send stuff to Facebook, Twitter, an email or SMS/MMS recipient, over Bluetooth (p.102) or Android Beam™ (p.101) to another device, or any number of other destinations added by other apps you have installed (sending an image to Google Goggles™ for analysis, or to your Facebook or Flickr photo albums, or as an attachment to an email for example).

Depending where you're sharing from, you'll either see the **Share** button in the Action bar or you may find sharing hidden among

further options when tapping the button. You can also often access sharing (and more) for an item by long-pressing it.

The destinations on offer will change depending on the type of file you're trying to share and whether or not the sites in question can handle it. You can share a photo with your Flickr account, of course, but if you select an MP3 to send, you won't see Flickr as an option.

Some of the sharing destinations available from the photo gallery. Different apps will share different kinds of items, so depending what you're sharing, your options will change.

apps & the android market

The Android Market

It's not just about the apps

Pretty much everything you can do with your Android™ device – from playing games or keeping up with Twitter, to remotely managing your business or editing Office documents – is achieved with an application or its related widget. As you'd expect, your tablet or phone will come pre-installed with enough features to make it usable – email, a web browser, contact lists, Google Maps™ and so on. But once you've mastered the built-in applications for these basic functions, you'll probably want to start installing third-party software for a whole range of other purposes.

This is where the Android Market™ comes in. It's your one-stop shop for almost all the Android software currently available (with the exception of the Android platform itself). Some of these apps are free, while others cost a few dollars. Once you decide on an app, so long as you have an Internet connection, you can have it downloaded and installed in a matter of seconds. The Market exists both as an app on your device and as a website you can peruse from your computer's web browser (or, indeed the Android browser). You can install apps directly from the **Market** app on your phone or tablet or log in to the website with your Google™ ID and have it push apps to your Android device from there.

Using the Market app

If you can't see the Market icon on your home screen, open up the app tray (see p.40). There'll be quite a few items there already, so swipe through the alphabetically arranged pages till you find the **Market** app. You may find yourself heading to the Market quite a lot so it's worth placing it on your home screen (see p.47).

Browsing for apps

Tapping on the **Market** app opens up a screen like the one to the left – an arrangement of panels advertising various "featured" apps and games that may or may not interest you. If you really want to, you can scroll down this page to see yet more adverts. Somewhere in the middle of the first screen, though, you'll see some smaller panels, labelled **Apps**, **Games**, **Books**, **Films** and (in the US, at least) **Music**. Tapping on any of these will take you to the relevant section of the Market, where you'll be faced with yet another page of advert panels. You may also see smaller green panels linking you to the other sections, along with links to Staff and Editors' Choices.

Can apps contain viruses?

Technically, no. When an app is installed it provides a list of "capabilities" to the operating system, basically a list of all the different functions it will need to access. Once installed, it's impossible for the app to do anything (such as using your phone to make calls or accessing your GPS location) that it hasn't declared in its capabilities. It's worth scrutinizing these permissions that you're granting the application, to make sure it's not asking to do anything you'd consider unnecessary (see p.233).

Tap the magnifying glass to perform a search, and the Menu button for settings and the **My Apps** section (where you can manage any apps you have installed). Just below this you'll see a horizontally scrollable list allowing you to switch views between Featured apps Top Paid, Top Free, and other charts. At the far left you'll find the Categories page, which lists everything in a slightly more deliberate manner (games, books, business, entertainment and so on).

You'd be forgiven for finding the Market app a tad labyrinthine. If you can't seem to find what you're after, simply

… That's better. Once you're off the main boulevard, the app market becomes a little more straightforward to navigate.

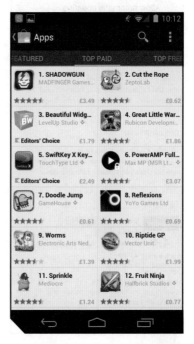

App windows just keep on going: this is less than half of the information you can wade through when perusing a potential download.

tap on the magnifying glass button at any point to bring up the keyboard and tap in a search term (or tap the microphone to perform a voice search). You won't get much info in your initial search results, just vital statistics, including the price and star rating (indicating its average popularity with users).

Installing a free app

Select any app and you'll be taken to a screen showing more detail: a few lines about what it does, how much of your device's memory it'll swallow up, and some screenshots that you can flick through or tap on for a closer look.

Will every Market app work with my device?

Probably not; the slowness of some manufacturers and mobile carriers to update to the latest version of Android™ means that users of older phones and tablets may find themselves left behind as newer apps come out that only support more recent releases of the platform. Some apps are just badly written or may not work properly with specific devices.

Something else you'll see now and again in the Android Market are apps that will only work if you have "root" access (which means hacking your phone – see p.125 for more).

Scrolling down the page, you'll see a handful of the most recent user reviews and ratings (depending on your device and Market version, you may have to look for a **Comments** tab, to take you to the reviews).

Tip: It's worth reading through a few of the user ratings as they'll often give you an indication of any problems the app may have on certain devices, or if it's just plain not worth bothering with.

Depending on your version of the Market app, you'll either see an **Install** or a **Download** button at the top of the screen. Tapping it takes you to a list of the services and functions the app will have access to. Give this list a quick once-over (see p.234 if you're not sure why) and tap **Accept & download** to install the app. Depending on the size of the file and your connection speed, it should saunter down the line in a few seconds. A small animated 🔼 will appear in the notification bar while it's downloading, changing to a tick ☑ once the app is installed. Dragging down the notification bar reveals a message confirming the installation. Tap on this message to set the app running if you want to try it

You only need to buy apps once

If you factory-reset your phone or tablet, switch to another one or have multiple Android devices, you won't have to pay more than once for your apps.

Log in to the Android Market with the same Google login that you use for your other devices, and you'll see all your previously downloaded apps in the Download section. From here you can reinstall any of your apps and those you've already paid for will be free.

out straight away. From there on, your new app can be found in the apps list (**Menu > My apps**) or can be placed on your home screen for easy access (see p.47). Occasionally an app will fail to install, usually because of a dropped wireless connection or a lack of space on your device. If that happens you'll see a notification to that effect, which you can tap to head back to the app's install screen so you can try again, although you may need to uninstall something else to free up some space first (see p.73).

Tip: Search apps directly from the Google Search bar, to avoid loading up the Market app.

Reviewing and rating your downloads

If at any point you want to check which apps you've downloaded, visit the Market, select the aforementioned **Menu > My apps** button, and you'll see a list of everything you currently have installed. Select one and scroll down till you see the opportunity to **Rate & Review** the app, where you can assign a star rating and make any comments you feel necessary. Comments appear almost

Reviews are usually a good way to get a feel for an app's usability and the speed with which its developer responds to any issues raised. Combined with the average star rating on an app's page, they can be an invaluable resource for narrowing your search.

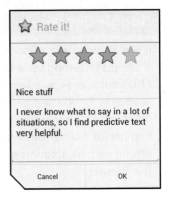

instantaneously in the Market alongside any other reviews. Earlier incarnations of the Market allowed you to clear your comments and ratings, but at the time of writing such a feature no longer exists. If you want to remove your comment you can tap on **Your Review** and edit it to delete or change the comment, but your star rating will remain until the end of time.

Installing a paid app

Sooner or later there'll be a paid-for app on the Market that you want to get your hands on. Many paid apps have a free trial version you can check out first, but if there's no such trial for this particular app, you have a slight buffer against wasting your money with the Market's (not exactly generous) fifteen-minute return policy (see p.73). Installing a paid app follows much the same process as for a free app, but with an extra few steps the first time around.

First, instead of the Install button at the bottom of the app's screen, you'll see a button with the price of the app in your local currency. Tapping this takes you to a Google Checkout™ screen which, on the first occasion, will ask you for your credit card details. You can enter these here or, preferably, take a minute or two out to set up a

Google Checkout

If you already have a Google account for Gmail™ (and you should – see p.34), setting up a Google Checkout only takes a minute. Checkout acts as an intermediary, handling your Android Market payments for you and passing these charges on to your credit or debit card. It's a free service that can be used with an increasing number of online retailers and means you don't have to keep setting up new accounts or handing your card details out all over the Internet.

Google Checkout account on your computer (see box above).

Once you've entered your details you'll see a message that the transaction is being authorized. The first time you use Checkout, your phone will ring at this stage (either your Android phone or your landline, whichever number you entered when you were filling in your details). There will be a friendly robot at the other end of the line who'll ask you some personal details in order to verify your purchase. Answer the robot's questions nicely. That's it! You can now download and start using the app. Meanwhile you'll get a confirmation email sent to your Gmail account.

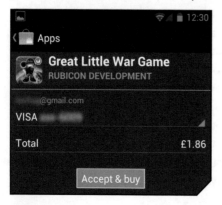

Once your details are registered with Google Checkout you'll never need to enter them again, and installing paid apps is pretty much the same speedy process as for free ones.

Can't find what you're looking for?

If an app is incompatible with your version of Android, or your particular phone or tablet, it won't even show up in a search from the Market app. It *will* show up on the market.android.com website but will still be flagged as incompatible and you won't be able to install it. Apps can also be restricted by region, with some only made available to certain territories. If you know an app is out there but can't find it, try searching some of the alternative sources listed on p.79, or try a Google search for the name of the app with ".apk" to find and install it manually (make sure you have malware protection if you go down this route, p.245).

Act fast for a refund

If, after checking it out, you decide that the new app isn't quite your cup of tea, you can return to its page (via the **Menu > My apps** in the Market app) within fifteen minutes of making your purchase, where you'll find an **Uninstall and refund** button near the top of the screen. A message sent to your Gmail account will provide confirmation once you're done.

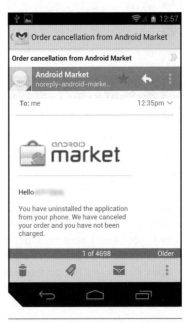

Updating apps

Looking at the **My Apps** section of the Market, you may notice

If you're quick, getting a refund is as easy as buying the app in the first place.

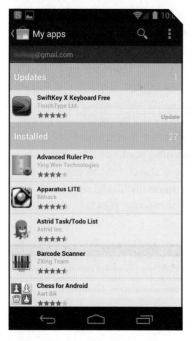

Apps update frequently to add new features, fix problems, or provide compatibility with new devices. Occasionally a fresh update will make an app that worked perfectly well before suddenly start encountering all kinds of issues. You can contact the developer directly via the Market or leave a review to politely draw their attention to the problem, or wait for others to do the same, and the problem will likely be resolved in the next update.

that updates are available for one or more of your apps (you'll also periodically receive messages in your notification bar to this effect, so don't feel that you have to keep checking). Just tap the **Update All** button at the bottom of the screen and they'll be queued to update without any need for further action. If you're running version 2.1 or lower, however, you'll have to manually select and update each app.

Tip: Manufacturers often neglect to properly associate bundled apps with their Android Market equivalents. Consequently they won't show up in your Android Market apps list, and may not receive updates. To remedy this, you can search for the app in the Market and install it to update your copy, and get future updates without having to check manually.

If you'd like newly installed apps automatically added to your home screen, head to **Menu > Settings** and check the **Auto-add**

shortcuts option. While you're there, check the **Update over Wi-Fi only** box to make sure your apps don't chew up your data allowance while updating.

Delve into **Market**'s settings page and you'll find an option to auto-update all apps by default. You can also set individual apps to auto-update on the app's page from **My Apps**.

> **Tip:** You may need to manually restart apps after they've updated, so if you have apps that are meant to run in the background – alarm clocks, schedulers or automation apps, for example – switch off auto-updates for these to avoid missing a scheduled event.

Browsing apps from your computer

Browsing the Market™ from the small screen of a phone or tablet is about as user-friendly as it could be under the circumstances, but for a less cramped experience, fire up your computer's web browser (or indeed, the browser on your Android™ device, if for some reason you don't get on with the Market app) and direct it to https://market.android.com. Once there, you'll find the same apps, games and widgets, but presented in a much friendlier interface.

The Market page is split vertically into two sections. The main centre panel presents a grid of featured apps while over on the left there are a couple of tabs – **Top Charts** and **Categories**, offering various subsections through which you can start browsing more deliberately.

> **Tip:** At the top of the left column you'll see (if you're logged in) an area informing you of the app's compatibility (or lack of it) with your devices.

Android Market

Search 🔍

Android Market › Education › NASAApp

NASA App
NASA

★★★★★ (1,535)

INSTALL

NASA App

OVERVIEW | USER REVIEWS (378) | WHAT'S NEW | PERMISSIONS

Users who viewed this also viewed

Satellite AR
ANALYTICAL GRAPHICS, INC.
★★★★★ (1,722)
Free

space junk
CASS EVERITT
★★★★★ (605)
Free

Space Images
JET PROPULSION LABORATORY
★★★★★ (123)
Free

NASAImages
ANDREI GOUMILEVSKI
★★★★★ (996)
Free

Users who installed this also installed

Hubble Center
VIRTUALNI ATELIER
★★★★★ (14)
Free

Solar System Explorer Lite
NEIL BURLOCK
★★★★★ (21)
Free

ISS Lookout
MARCELO MARTIM MARQUES
★★★★★ (30)
Free

NASA Images Archive
NGO
★★★★★ (362)
Free

Description

The Official NASA App for Android
Come explore with NASA and discover a huge collection of stunning images, videos, mission information, news, NASA TV and featured content with the new NASA App for Android.

Features:
- Thousands of images from NASA IOTD, APOD and NASAImages.org
- On demand NASA Videos from around the agency
- Current NASA Mission Information
- Launch Information & Countdown clocks

MORE

Visit Developer's Website ›

App Screenshots

User Reviews

5 star ▮ 1246
4 star ▮ 376
3 star ▮ 152
2 star ▮ 71
1 star ▮ 124

Average rating:

4.3

★★★★★
1,535

AMAZING. Video is great. If it doesn't work then you need to upgrade your ...
★★★★★ by **natalie** – July 31, 2011

AMAZING. Video is great. If it doesn't work then you need to upgrade your phone. It's beautiful, but so sad--our poor space program....

Read All Reviews ›

+1 | 268
🐦 Tweet

ABOUT THIS APP

RATING:
★★★★★
(1,535)

UPDATED:
July 15, 2011

CURRENT VERSION:
1.22

REQUIRES ANDROID:
2.1 and up

CATEGORY:
Education

INSTALLS:
500,000 - 1,000,000

last 30 days

SIZE:
3.9M

PRICE:
Free

CONTENT RATING:
Low Maturity

It's not just for apps, by the way

As well as apps and games, you can now use the Android Market to rent movies, download eBooks and listen to music. Head to the Media section of this book (p.183) to find out more.

Click on any app's icon to head to its main page. Here's where the Market's web interface really has the edge over the mobile version, as you can see pretty much everything you need to know at a glance.

Installing from the web interface

The **Install** button can be found in the top left of the screen, in the square panel where the title is. If an app needs paying for it'll have a button showing the price instead. Upon clicking the button, you'll be taken to a checkout window. Select the device you'd like to install to (if you have more than one) from the drop-down menu in the top half of the window. The bottom half outlines which permissions the app requires. Give this a quick once-over to make sure there's nothing dodgy going on (see p.233) and then hit the **Install** button, or **Cancel** if you've changed your mind. For a paid app, press the **Continue** button and follow the instructions.

Assuming your device is switched on with a working Internet connection, the app will be downloaded, installed and ready to use next time you connect. If not, it'll download at the next available opportunity.

The Android Market web page (left) lets you browse from the relative comfort of a full-size computer screen. Most of the important information can be accessed from the same screen, and everything else is available from the tabbed centre panel.

Can Android apps run on my PC?

In a roundabout kind of a way, yes: you can download and install the PC applications **BlueStacks** on your computer and run apps and games from there. You may encounter a slight performance lag as both programs need to do a lot of behind-the-scenes legwork to provide a suitable environment for the apps to run. BlueStacks currently comes packaged with a number of popular apps and games, leaving a few empty slots for you to add a few of your own. You can also "push" apps from your Android phone or tablet to your PC using the **BlueStacks Cloud Connect** app (available from the Market). Open BlueStacks from the desktop gadget and select **Get More Apps**. You'll then need to log in to Facebook and allow BlueStacks access to your data, after which you'll get a PIN to enter into the app.

▶ **BlueStacks** bluestacks.com

Although still in its infancy, BlueStacks runs Android apps and games on your desktop PC with ease.

Other ways to browse & install apps

As you'd expect, there's a whole universe of Android™-devoted websites out there, and many of the geeky tech review sites have regular roundups of the best apps.

▶ **Recombu** recombu.com/apps/android

▶ **Gizmodo** gizmodo.com/tag/androidapps

▶ **Androinica** androinica.com/category/google-android-applications

▶ **Lifehacker** lifehacker.com/tag/android

▶ **Android And Me** androidandme.com/category/applications

▶ **Android Police** androidpolice.com/topics/applications-games

▶ **Android Central** androidcentral.com/reviews/software-reviews

Most sites link to the Android Market™, others publish a QR code along with the review (see p.256). Simply point your device's camera at your computer monitor using the **Barcode Scanner** app, and you'll quickly be taken to the install screen.

AppBrain

AppBrain is a popular alternative to the Android Market: to use it you'll need to install and run the **AppBrain App Market** and **Fast Web Installer** apps (search for "AppBrain" in the Market). Log in to the **AppBrain App Market** app using your Google™, Facebook or Twitter account settings and give your approval for the Fast Web Installer to do its thing. The first time you run the app it's probably worth tapping the **Manage and Sync** button and hitting **Sync** (bottom left) so that the website knows what you already have installed.

Once you've set this up you can browse to appbrain.com on your computer and start installing apps. The site is very straightforward:

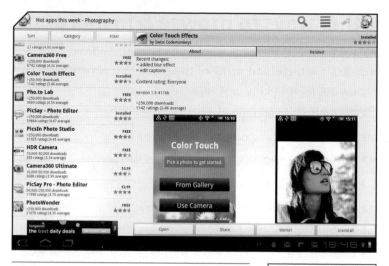

AppBrain was ahead of the curve when it came to providing a web interface for installing and uninstalling apps to Android devices. Google's own Android Market web interface has finally caught up, but AppBrain still has many loyal followers and still seems to be the only service that enables you to uninstall via the web.

you can browse apps by category, get recommendations, see what's new and so on. When you find an app you like the look of, click the **Install** button. A pop-up window will show the app's permissions (see p.233) for approval. Click **Install this app** and it will instantly download and install on your device. You can also uninstall apps via the website from the **My Apps** section. From here you can click **Cancel Install** for apps that you've installed via the site, or **Uninstall** for other apps. Firing up AppBrain on your device, tap **Manage my apps > Sync with AppBrain** to see a list of pending uninstalls. Tapping **Perform installs** will set any queued installs or uninstalls in action.

For installing paid-for apps you'll still need to hop over to your phone or tablet to confirm your payment credentials, but for browsing and installing free apps it doesn't get any simpler.

> **Tip**: When looking at paid apps, shop around between download sites and the official Market, as you may find the app you want is available cheaper or even for free.

Amazon

The Amazon Appstore (only available in the US at the time of writing), works in much the same way as the Android Market or AppBrain, with the added convenience that you may already have a credit card set up with Amazon. You'll need to install Amazon's own

Appstore app in order to use the service; head to the Appstore website, try to install something and you'll be prompted for your email address. A link to the Appstore app will then be emailed to you. Before you can install it, you'll need to head to **Settings > Security** and make sure the **Unknown sources** box is ticked. One of the best things about Amazon's Appstore is their daily free or discounted app offers – worth keeping an eye on for some great deals.

▶ **Amazon Appstore** http://
goo.gl/LOYR1

GetJar go one step further than Amazon's free app of the day with their "Gold" section which offers up a bucketload of paid-apps for free at any given point in time.

Other non-Market app sources

Many of the sites below have their own Market-like apps that allow you to install stuff in pretty much the same way as the official one.

▶ **GetJar Gold** getjar.com/gold

▶ **Soc.io Mall** mall.soc.io

▶ **Handango** handango.com

▶ **SlideMe** slideme.org

▶ **Phoload** phoload.com

Alternatively, you can download .apk files directly from the web – either to your computer and then dumped to your Android device's SD card, or more directly through your device's web browser – and install them with the help of an app such as **Easy Installer** from the Market. Most file manager apps (see p.108) also provide a facility for installing apps from the SD card, so if you have one installed already, dig deep into its features to see if it has that capability.

calls & messages

Calls & messages

Keeping in touch

Android™ started out first and foremost as a mobile phone platform, so as you'd expect, telephony and messaging are well developed, fully integrated and easy to use.

Phone calls and text messages can be accessed from their apps (**Phone** and **Messaging**, usually residing in the Docker); by tapping a related item from the notification bar; or from your contacts list (or any other app that displays your contacts' photos) by tapping on a person's picture and selecting the appropriate contact method.

A typical contact page from tapping a picture in your **People** list, Phone contacts or any other app that draws contact data from your phone (**Gmail**™ for example). The top row will display icons for as many different contact types as you have for that person.

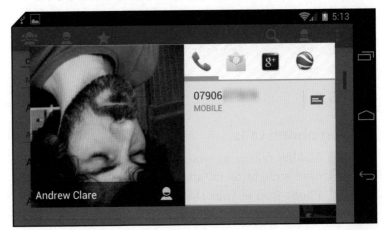

The three tabs in the Phone app (top left) switch you between the dialler (shown here), call history and contacts list. As with the rest of the Ice Cream Sandwich launcher, you can tap the tabs or just swipe left and right to move between these screens. While making a call (right), the Action bar displays various options, including speaker/microphone mute, hold and **Add call** (for making conference calls, see p.88).

Calls

Making phone calls

 Making calls is as straightforward as you'd expect. To head straight to the dialler to make a call, select the **Phone** icon in your Docker. Tap in the number to call and press the call button at the bottom of the screen.

For hands-free calling you can also use the Voice Dialler; select it from the app tray and speak the number or name of the person you'd like to call.

The Voice Dialler app (above) can be found tucked away in your app tray. Invaluable for making calls while driving, skateboarding or prize fighting.

Receiving phone calls

Receiving calls is much the same as on any other phone. Depending on your model, you may need to swipe a bar or unlock icon in one direction or another. The stock Android™ 4.0 dialler will display a large image of your caller (or a placeholder if your caller doesn't have a contact image in your list) and a not-immediately-obvious button in the middle. It's not until you tap and hold this button that your options become clear: drag left to reject the call, right to answer it, or drag up to select and send a "Quick response" – basically a canned response via text messaging. Quick responses are editable by tapping the menu button from within the dialler and going to **Settings > Quick responses**. There are a handful available, which you can tap on to change to anything you like.

The stock call answering interface in Android 4.0. Branded models may well have their own system, so you may need to refer to your phone's manual.

Can I make phone calls while on the Internet?

If you have access to a Wi-Fi Internet connection, then yes you can. For example, if a friend called you up and needed directions, you could put them on speakerphone while you looked up their location in Google Maps™, talking them through the rest of their journey. Android also supports simultaneous voice and data using the 3G, 4G and EDGE data networks, but, as usual, whether your handset and/or carrier does may be another matter.

Multiple calls & conference calls

Pressing the **Add call** 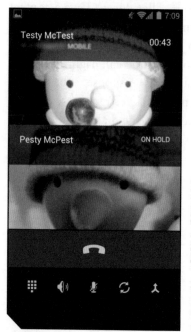 button during an existing call lets you select and dial another number while your current contact is placed on hold. To switch between calls, press the **Swap** button. You'll notice that the Add call button has now transformed into a new **Merge calls** button. If your carrier allows it, you can host a conference call by tapping this and talking to everyone at the same time.

> **Tip:** You can quickly hang up from a call by pressing your phone's **power** button.

Android 4.0 has built-in support for conference calls. Before you get too excited though, contact your mobile phone carrier to see if they'll allow it under your current tariff.

Voicemail

To pick up your voicemail, tap on the notification from the notification bar, or fire up the dialler and long-press the number 1 key (if you haven't entered your voicemail number in the **Settings > Voicemail settings > Voicemail number**, you'll need to do that first, or manually dial it for the rest of your life).

Google™ have built the framework for visual voicemail into Ice Cream Sandwich but there's no native app in the stock version. This is something that may or may not be bundled with your phone, depending on the carrier and whether or not they've integrated it. If your carrier has provided a visual voicemail app, you'll be able to jump through your messages, fast forward or slow them down, receive text transcripts, and generally engage with your voice messages in a much more useful manner.

There are a number of third-party visual voicemail apps and services in the Market, including **YouMail** and **Visual Voicemail Plus**, but before you try either of these, try **Google Voice**™ (see p.90).

> **Tip:** You can set a different ringtone for a specific contact, or send their calls straight to voicemail. Select the contact from the **People** app (p.136) and tap the **Menu** button for these, and other options.

Making calls over the Internet

As well as regular calls and messages, you can use your Android device to talk and make video calls over your Internet or (carrier permitting) your data connection. There's a range of VoIP (Voice over Internet Protocol) clients and services available from the Android Market™, usually offering some combination of pure Internet calling, video chat and cheap calls to landlines in other countries.

Google Voice Free

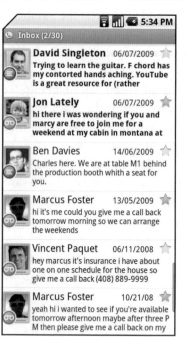

Google Voice gives you complete control of all your phone numbers at once. The only thing it's missing is video chat. Although available outside the US, at the time of writing many of Google Voice's features still only work from a US-based phone number.

Add the **Google Voice Callback** app, configure it correctly and you should be able to use Google Voice to make free phone calls.

Google Voice is a multi-faceted VoIP telephone service. It gives you a single phone number which you can use forever, irrespective of where you live or which phone services you use. You can also use it to replace your existing voicemail service for the ability to listen to your voicemail online, and receive visual voicemail text or email transcripts of your messages. It also acts as a switchboard for all your phones, letting you route calls to any of your other numbers (landline, mobile, office etc), or ring multiple numbers at the same time. The service also offers free text messaging and cheap international calls.

Skype Free   3.7

Whichever VoIP service you choose to go with will depend largely on which service most of your contacts are using. Like it or not, Skype is a ubiquitous option. The Skype app for Android lets you make free voice and video calls to other Skype users, as well as offering cheap international phone calls.

Fring Free   3.8

Fring, like Skype, offers video chat, voice calls and live text chat. You can call other Fring users around the world for free. Where Fring has the edge over other services is that it provides group video chat for up to four users. It's also available on iPhone and iPad so there's no obstacle to staying in

touch with your iFriends (apart from their thinly veiled jealousy of your Android device and complete denial of its superiority).

The best of the rest

Other great VoIP and cheap-call apps worth checking out include **GrooVe IP**, **Tango**, **Viber**, **Rebtel** and **TiKL**, which turns your phone or tablet into a push-to-talk style walkie-talkie.

Messages

Text messaging

To send a text (SMS) message, start up the **Messaging** app, and select a thread or message to reply to, or start a new message by tapping the ▤ button in the Action bar. As you begin to type the name of your recipient into the **To** field, an auto-search will pull in names for you to select from. Next, tap into the **Type message** field and start typing (or hit the microphone 🎤 key to dictate, see p.59). You can also switch directly to a phone call by tapping the 📞 icon in the Action bar, or tap 📎 to attach a range of files and send your message as an MMS (multimedia messaging service) if your contract allows it.

If you're starting out from the **People** app or accessing your contacts list from anywhere else on the phone, tapping the ▤ icon on any contact will instigate a new text message to

MMS messages can be sent to another phone or to an email address. You can format text and attach a dizzying range of multimedia file types. Larger messages may require a 3G/4G data connection and may incur an extra charge above your normal tariff.

that person. You can augment or swap out the messaging app with a host of SMS apps from the Market, some standout examples being **Go SMS Pro** and **ChompSMS**.

Handcent SMS Free   `4.3`

Powerful, customizable SMS/MMS replacement brimming over with features including: password security; a contact locator (find your friends using GPS); group sending; per-contact notification, ringtone and vibrate settings; themes, skins and font packs; message scheduling; backup to a Handcent Online account; integrated blacklist for blocking spam; and heaps more.

HeyWire Free   `4.3`

More a complete messaging hub than an SMS replacement, HeyWire lets you send text messages for free to 45 countries worldwide. It also has Facebook and Twitter chat, useful if you want all of those things in the same place. Similar free worldwide text services can be found in the Market from **JaxtrSMS, WhatsApp** and **TextPlus**.

IRC chat

Most of the main IM chat clients have an Android™ app of their own, but unless everyone you chat with uses the same network, it's worth checking out a multi-protocol instant messenger, such as **eBuddy Messenger** or **imo**. These hook up with multiple accounts (MSN, Facebook, Yahoo!, AIM, ICQ, Google Talk™, MySpace and more), providing a one-stop solution to all your chat needs. As always, these apps are all available from the Android Market™.

Trillian Free · 4.0

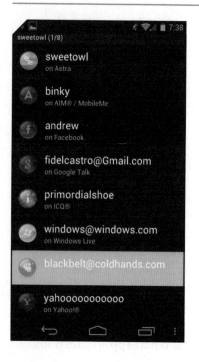

Trillian lets you create one account to consolidate all your online chat identities. It works with all the major networks and you can even read and post updates to Facebook and Twitter. Adding contacts is easy and if friends are on multiple networks themselves you can combine these into a single contact.

getting technical

Getting technical
Taking a look under the hood

So we've covered all the basic stuff: how to operate your phone or tablet and how to use apps and widgets. Now it's time to roll up your sleeves and dig a little deeper into your device.

In this section you'll find out how to control your mobile data usage (p.112), disable crapware (p.116), extend battery life (p.117) and discover a whole world of other customizations and tweaks that'll help you use your Android™ device to its full potential.

Connections

Before we go any further, let's take a quick look at the different types of connection available from your device. As well as USB, Wi-Fi, and 3G/4G/EDGE connections, your Android device may be equipped with NFC and Bluetooth chips, and be able to connect using the new Wi-Fi direct protocol.

Wi-Fi

 If you're within range of a Wi-Fi network, as found in homes, offices and cafés, and across some entire cities, this will probably be your preferred choice of Internet connection. It's fast and generally reliable, easy to set up, and often free.

If you don't see the network you're looking for, tap **Scan** at the bottom of the screen to re-search the airwaves. To connect to a hidden network you have access details for, tap **Add network** and enter the relevant info.

Setting up a Wi-Fi connection to a home router or wireless hotspot is normally a fairly simple procedure. When you first switch on your device and attempt to sign in to or create a Google™ account, you'll be taken straight to the Wi-Fi settings screen. At any other time you can find it from **Settings > Wi-Fi**. A toggle at the top of the screen lets you switch Wi-Fi on or off. Below that you'll see a list of all the wireless networks currently within range. Tap on the one you'd like to connect to. If it's a private network (such as your home network) enter the security password when prompted, and you're good to go. Android will remember login details for any networks you connect to and will be able to connect to them again whenever in range.

Tip: If your passcode is a zillion characters long and full of odd glyphs that the Android keyboard doesn't even have, your best bet is to get it onto your device as a text file over a USB connection, or via email (which you can pick up with a mobile data connection, if you have one), and to then copy and paste when prompted for it.

Wi-Fi Analyzer Free 4.6

Shows all the nearby Wi-Fi access points and helps you improve your own Wi-Fi signal at home by finding a less crowded channel for your wireless router. If you're out and about, you can use it to find out which hotspot has the strongest signal and least traffic. With the free add-on Wi-Fi Connector, it even lets you switch connections from within the app. Wi-Fi Analyzer offers a number of different views of the state of your airwaves, including a channel graph (below) and a signal strength meter, useful for finding hot and cold spots around the home.

Dodgy connection? Get a quick window on who's battling you for the airwaves. Wi-Fi Analyzer presents its findings in a number of ways, focusing on either signal strength or channel usage. Keep a network cable handy if you're remotely configuring a Wi-Fi router, just in case something goes awry and you need to change your settings back (it'll be much easier to reconnect to your router with a physical cable if you've messed up the Wi-Fi settings).

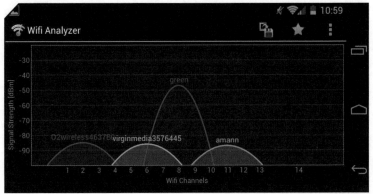

Wi-Fi file explorer Pro $0.99 (£0.59) 4.6

Simple to use but feature-rich app which opens up your device's Wi-Fi file management possibilities. No more messing around with USB cables; you can browse, transfer, edit and even stream files to and from your Android device via a web browser and Wi-Fi connection.

3G/4G/EDGE

 When out and about, away from any accessible Wi-Fi networks – walking down the street, say, or sitting on a bus – you'll need to use a mobile network connection (3G/4G or EDGE) to access the web. A 3G connection won't be as impressive as that experienced over Wi-Fi, but it should be adequate for most needs (see box opposite for more details).

Another factor when out and about is your own speed of travel. If you're on a train or in a moving car, you might find that your phone's connection speed is slower than when you're stationary. This is because it's having to accommodate a constantly shifting relationship to the nearby signal masts that it's connecting with, making it hard to maintain a constant and coherent stream of data to and from the Internet.

Tip: If your mobile 3G/4G Internet stops working, you could try enabling and then disabling Airplane mode (press and hold your phone or tablet's **power** button for a menu which includes this option). Or, if that doesn't work, reboot your device.

2G, 3G, 4G ... what's all that about?

Over time, the technology used to transmit and receive calls and data from mobile phones has improved, allowing greater range and speed. Of the network technologies widely available at present, 4G (fourth generation) is the most advanced, allowing Internet access at speeds comparable to home broadband connections (at least when stationary). Your device will automatically use the best available connection that your SIM supports, switching to 3G or 2G networks when 4G isn't an option locally.

2G, GPRS (2.5G), and EDGE (2.75G) offer lower speeds more akin to an old-fashioned dial-up connection. One advantage these older technologies have over 3G and 4G is that they're much more economical in terms of battery power. If keeping your device running a bit longer is a priority, you can make sure it doesn't attempt 3G or 4G connections by heading to **Settings > Wireless & networks > Mobile networks** and selecting **Use only 2G networks**.

NFC (near-field communication)

If your device has a built-in NFC chip, you can share items and information with other NFC-enabled Android devices using a new protocol called Android Beam™.

Once you have the content you'd like to share on screen, it's as simple as placing the two devices back-to-back; a window will pop up describing what you're about to share. Tap the window and your share is done. Among other things, you can share contacts, web pages, YouTube™ videos, map directions and apps. What you're actually sharing in most instances is a simple web link or a very small amount of data (such as a contact or position in a game). NFC transfer speeds are pretty slow, so it's not really set up for sending actual audio files, photos or videos. For sending large files directly between devices you can use Bluetooth, USB or Wi-Fi Direct.

NFC can also be used to make payments at a growing number of real-life retail outlets using the **Google Wallet**™ app.

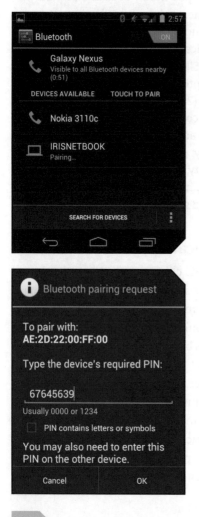

Bluetooth

Bluetooth is a short-range wireless technology found in headsets, external keyboards and mice, hands-free kits for cars, laptops and other portable devices, and game consoles. Your device can connect with other Bluetooth devices within an eight metre or so radius.

To instigate a Bluetooth connection you first have to make sure that Bluetooth is switched on, either from the **Power Control** widget (p.118) or by toggling the Bluetooth **on/off** switch in **Settings**. The first time you connect a new Bluetooth device you'll need to "pair" it with your Android device (subsequent connections will be automatic). You also need to make your device visible to other Bluetooth devices. Go to

Settings > Bluetooth. Your device will start scanning for nearby Bluetooth-enabled devices (make sure that any devices you want to connect to are switched on and have been set to discoverable).

Once your Bluetooth devices are visible in the **Devices Available** list, tap the item you'd like to pair with and enter the appropriate PIN (your Bluetooth device will either provide one, or have a default one listed in its manual – if you can't find it, try 0000 or 1234 as these are pretty common). If all goes well you'll now be paired with the device.

To connect to this device again at a later time, head back to **Settings > Bluetooth** and tap it from the list of paired devices.

Bluetooth file transfer Free   4.3

Handy little app that lets you securely browse, explore, share and manage files and contacts on nearby Bluetooth devices. It has a fast, customizable file browser and can even search within and create compressed (Zip, Tar and GZip) archives.

Torque/Torque Pro Free/$4.95 (£2.95)   4.8

If your car has OBD (on-board diagnostics), Torque will connect to it via Bluetooth and glean all sorts of valuable diagnostic information about performance, faults and sensor readings. Set up a dashboard with widgets and gauges and see what your car is doing in real time.

TouchDAW Free/$4.75 (£2.99)   4.6

Another stellar example of what's possible with a Bluetooth device. TouchDAW is a powerful MIDI controller for use with audio workstations. It can function as a keyboard, control surface, launchpad and X/Y controller. You can even hook up your device's internal sensors to trigger MIDI events.

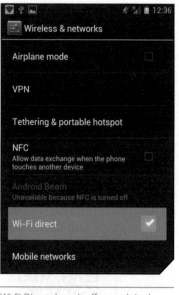

Wi-Fi Direct doesn't offer much in the way of settings: a simple check-box pretty much covers your options. Whether you can persuade your other devices to cooperate is another matter.

Wi-Fi Direct

Wi-Fi Direct is a new Wi-Fi protocol which enables direct, secure ad hoc connections to other Wi-Fi devices without the need to join a traditional home, office or hotspot network. It's easy to set up and is already being hailed in some quarters as a replacement for Bluetooth. When your device is enabled as a Wi-Fi Direct host, other Wi-Fi devices (or "clients") can connect to it and transfer data. You can use it to move photos, stream media between devices, share an Internet connection and more. Enabling Wi-Fi

Direct is as simple as hitting up the **Settings** menu and tapping **More…** under the **Wireless and networks** section. Then simply select the Wi-Fi Direct option and you're away.

USB

You can use a USB cable to transfer music, pictures and other files between your Android device and a Windows computer. The standard connection type is MTP mode, which is supported from Vista onwards. If you're connecting to XP or an earlier operating system, tap on the connection notification and switch to **Camera mode**.

Your device's SD storage (p.107) will become mounted and appear as a drive in your computer's file browser, allowing you to move files around as you would for any other peripheral storage device.

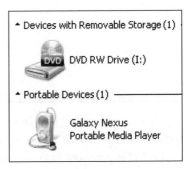

Depending whether you've set your USB connection to MTP (above) or Camera mode (below), your device will show up as a media player or camera in Windows. The same files and folders will be accessible no matter which you choose.

Tip: If you take photos or write any other kind of files to your device's SD storage, your Windows machine won't see these new items while the two devices remain connected. To trick Windows into taking a fresh look at your files without having to disconnect and reconnect the USB cable, you can switch between MTP and Camera modes on the fly (effectively resetting the connection) and your new files will become usable.

Mac users require a couple of extra steps to set up file transfer to and from an Android device, but once you're up and running it's as straightforward as working with any other external file storage.

If you're tethering your Internet connection over USB , you'll have to switch this off before you can make a file transfer.

The MTP protocol isn't supported by Mac OS X, but you can download and install the free **Android File Transfer**™ application from android.com/filetransfer. Double-click the app to use it for the first time (after which it'll run automatically whenever you connect your Android device). Android File Transfer lets you drag files around between your devices in much the same way as you would with a standard Finder window.

Dropbox Free   4.7

A free service that synchronizes your stuff across all of your devices. Just drag an item into your dropbox and it'll upload in the background, automatically saving to your other devices and to Dropbox's own servers. A great tool for backing up and transferring files.

Storage and memory

SD? RAM? ROM? What's the difference?

ROM (Read Only Memory) is used to store your phone or tablet's operating system and any apps you have installed (hence you'll often see the operating system itself referred to as a ROM, see p.257). These are loaded into the RAM (Random Access Memory) when needed. Together these two memory chips make up part of the computer that runs everything on your device.

SD (Secure Digital) is flash memory, similar to the kind found in digital camera memory cards and USB memory sticks (the card itself is often integrated into the device, although some units provide an SD expansion slot for removable storage). You can store your photos, backups, music and other files on it as you would on a

computer's hard drive. It's much smaller than a hard drive, less power hungry and less likely to break if the device is dropped. When your device is connected to your computer via USB, you can browse the files stored on the internal SD card and move music, photos and other files between the two devices.

If you intend to make any use of your device's storage for

To see a breakdown of your available storage, head to **Settings** > **Storage**. Tap any sub-category (for example, Apps, seen here) for a better handle on what's eating up all your space.

photos, media, or any other kind of files, you'll find a good file manager is pretty much an essential tool. Luckily, there are some great free options out there, including this one...

ES File Explorer Free   4.7

A veritable Swiss Army knife of file management, ES File Explorer not only lets you explore and edit the files stored on your Android™ device, but can also connect via Wi-Fi and Bluetooth to provide access to files on your home computer and other devices, local area network, and the Internet. It has a built-in FTP client and cloud storage client with support for Dropbox, SugarSync and Box.com. As if all this wasn't enough, it also includes

file viewers, a text editor, compressed and encrypted file support, an application manager and task killer. Another great file manager worth a test drive is **Astro**.

The unlabelled buttons aren't especially intuitive, but mess around with them for a bit and you'll figure out what everything does.

DiskUsage Free

4.6

Gives you a complete visual overview of your device's storage. Zoom in and out and tap items to view them. An invaluable tool for hunting down space-hogging apps and files.

Moving apps to your SD memory

Devices with SD cards let you move some of your apps to run from the SD instead of the phone's operating memory. You may find yourself wanting to do this at some point if you have a lot of apps installed and start to run out of space for new ones, though this is less of an issue with the larger operating memory of newer devices.

From **Settings > Apps** tap on an app and, if it's available, you'll see an option to move it to the SD card. A small part of it (a "stub") will remain in the device's RAM so that the app can continue to function properly, and any widgets that are part of the app will no longer work. The capability for an app to be moved to the SD card has to be included in the installer by its developer, so may not be available in all apps you install.

Apps2SD Free   4.6

It can be a drag manually checking every single installed app to see if moving to SD is possible. Apps2SD rounds up all your movable apps into one place and lets you decide which ones to move. It also gives you a one-click option to migrate them all at the same time.

Internet

Tethering

Tethering enables you to share your phone or tablet's Internet connection with another device over USB or Bluetooth. Assuming you have a working Wi-Fi or data connection, Android™ supports both USB tethering and the ability to set up your device as a wireless hotspot. So, for example, if you needed to get your laptop online in a situation where no Wi-Fi signal was available, you could connect your phone or tablet and employ it as a 3G modem.

First, make sure your device is connected to your computer with the USB cable, or that it's paired via Bluetooth (p.102). Now head to **Settings > Wireless & networks > More... > Tethering & portable hotspot**. From here, you can switch on **USB tethering** or **Bluetooth tethering**. That's all you need to do; Android will handle the rest.

> **Tip:** Manufacturers and mobile carriers often disable tethering on their devices. If your phone or tablet is crippled in this way you can usually enable tethering without rooting. Look for **PdaNet, ClockworkMod Tether** or **EasyTether** on the Android Market™.

Portable Wi-Fi hotspot

Setting up your phone or tablet to act as a wireless hotspot is almost as easy as tethering. From the same screen, select the **Portable Wi-Fi hotspot** check-box. After a few seconds you should see "Android AP" as an available Wi-Fi connection on your computer. Back on your device, select **Configure Wi-Fi hotspot** and make sure the **Show password** box is checked. You can enter this password into your computer when connecting to the hotspot. The Configure screen is also where you can change the name (SSID) of the connection and the security type (although unless your computer doesn't support it, just leave this as WPA2 PSK).

Configure Wi-Fi hotspot

Network SSID
AndroidAP

Security
WPA2 PSK

Password
2b133b0454d5
The password must contain at least 8 characters.

☑ Show password

Cancel   Save

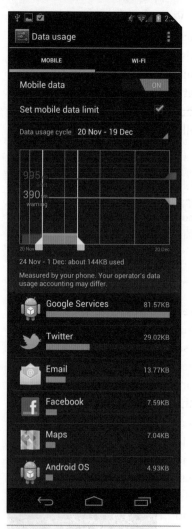

Running into the red every month? The data usage monitor not only keeps track of your mobile Internet usage, it can also set limits so that you don't go over.

Data usage monitor

If your mobile data provider (or Internet service provider) only allows you a fixed monthly data allowance, or charges extra for anything over a set data limit, you'll find the data usage monitor one of the most useful features added to Ice Cream Sandwich. Tap your way over to **Settings > Wireless & networks > Data usage** to take a look at it.

The graph at the top shows your data usage over time. The orange horizontal line can be dragged up & down to set a warning when your monthly usage goes over a certain point. Select the **Set mobile data limit** check-box and a similarly adjustable red line appears. Mobile data will be stopped dead in its tracks once this limit is reached.

Above this, you can tap to adjust the **Data usage cycle** and set it to wherever your monthly billing cycle starts and ends.

Data usage page for Google Maps™. Interesting to see how much background data is used (the purple area). Some apps will lose basic functionality if this is switched off completely (see below), so if possible, your first line of attack should be to visit the app's actual settings page (a handy button is provided) and adjust any relevant setting there.

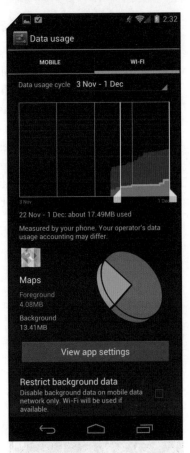

The bottom half of the screen shows a list of which apps and services have been using the most data. Tapping on any of these takes you to a dedicated page for the app where you'll be able to see how much mobile and Wi-Fi data it's used over time (Wi-Fi display can be switched on from the **Menu** button). You can't set a separate data limit for each app, but you can individually restrict each app's background data usage, so that, for example, it'll only update in the background when there's a Wi-Fi signal present. To do this, just scroll down the app's page and check the **Restrict background data** box.

Tip: Your data provider's records may differ from your own. To err on the side of caution, set yourself a slightly lower limit than your allocated monthly allowance to avoid being charged.

Performance tweaks

One of the Android™ platform's strengths is its ability to multitask like a computer – to have a whole bunch of stuff happening in the background while doing something else in the foreground.

Android manages applications in such a way that if a running app isn't currently in use it lies dormant in the device's memory (RAM) without using system resources. If memory is needed for some other task, Android will simply free up as much as it needs by quitting long-dormant apps. It also pre-loads certain background apps and functions into memory, but for the most part you can ignore these.

You don't need one of these nearly as much as you might think.

Task killers

A task killer is a small app that you can use to selectively close (kill) other apps on your phone or tablet. If you have one of these installed, you'll be able to see that a surprising number of applications seem to be running in the background and it's tempting to routinely kill as many of these as you can in the hope of making your phone run a bit faster. In fact, it costs the phone as much power to hold "nothing" in its memory as it does to hold actual data, so using one of these programs to constantly kill apps in an

For the rare instances when you may need one, Android has its own task killer built into each app's info page.

App info

Astrid Tasks
Version 3.8.5.1

Force stop | Uninstall

STORAGE

Total	3.95MB
App	3.90MB
USB storage app	0.00B
Data	48.00KB
USB storage data	0.00B

Clear data

attempt to reclaim resources is largely a waste of time, and may actually be causing you more problems than it's solving – you may miss notifications, alarms or email updates, or be forcing the device to reload regularly used apps back into memory all the time instead of just leaving them dormant for when they're needed.

But while task killers aren't really necessary for managing memory usage, they do have their uses. If you have a program that's crashed or isn't behaving properly, an app like **Advanced Task Killer** will give you a handy shortcut to force quit and reload it, which usually solves the problem. The same thing can be achieved, however, by going to **Settings > Apps** and scrolling down to the app you'd like to quit. Selecting the app takes you to a detailed screen with all kinds of information, including a button to **Force stop** and one to **Uninstall** the app altogether.

Certain apps may continually run in the background unless you deliberately quit them. If there's an app that you're convinced is running your battery into the ground, you can use a task killer to regularly nip it in the bud and see if things improve. Some apps, such as unnecessary bloatware loaded onto your device by the manufacturer or provider, will just start right up again, so force-closing them is largely fruitless. While you may not be able to uninstall these apps, you can disable them to prevent them loading in the first place (see below).

Several resource meter apps on the Market™, such as **SystemPanel**, combine a task killer with the ability to monitor individual apps' long-term resource usage. Handy for troubleshooting a slow or laggy device.

> **Tip**: Some task killers will automatically ignore Android's system apps, but if yours doesn't, you should add the following apps to your ignore list. These, plus anything starting with ".com", need to be allowed to run so that your phone behaves itself properly: Alarm clock, Bluetooth share, Browser, Calendar, Gmail™, Market, Messaging, My uploads, Package installer, Settings, Voice dialler and Voice search.

App disabling

Most of the apps built into your device's system are there for a good reason, and many are so much a part of that system that it's easy to forget they're even apps – the keyboard or dialler for example – but often manufacturers will throw in a bunch of unnecessary branded apps that you don't need and can't get rid of. Android 4.0 lets you disable these apps to prevent them running. Go to **Settings > apps** and select an offending item. At the top of the app's page, next to the **Force stop** button, you'll find the **Disable** button (for apps you've installed yourself there would be an **Uninstall** button in its place). Press the button, end of story.

Battery

With all the features loaded onto your device – a bright screen, apps and games, 4G, Wi-Fi, Bluetooth and GPS, to name but a few – your battery can have a tough time keeping up. When you first start using a new phone or tablet, battery life can be pretty short, but after it's been "broken in" with a few charge cycles its capacity will improve greatly. Here are a few tips on how to extend it even further:

▸ **Keep it cool:** Your battery will not hold its charge so well in high temperatures, so keeping your device out of your pocket and away from other heat sources will help.

▸ **Switch off 4G:** It's nice and fast but it's a real power hog. If you're not using it, turn it off. Check out **APNdroid**, a useful little app that lets you toggle it on and off without cutting your basic call signal.

Head to **Settings > Battery** for a full rundown of what's running it down. Tapping on an item in the list gives you a bit more detail, including advice for how to limit the particular app's power consumption.

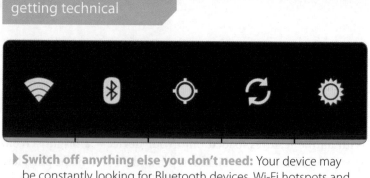

▶ **Switch off anything else you don't need:** Your device may be constantly looking for Bluetooth devices, Wi-Fi hotspots and GPS position. Placing the inbuilt **Power control** widget (above) on one of your home screens makes it easier to toggle these on and off as required. It also lets you switch your screen to a lower or automatic brightness level, another great way to save battery juice (see below).

▶ **Update less often:** Just because your phone or tablet can pull in Facebook updates, weather reports, stocks and shares, and so forth on a minute-by-minute basis, doesn't mean it should. From **Settings > Accounts & sync** you can select each service you subscribe to and adjust the frequency it updates.

▶ **Tweak your screen settings:** Changing the time-out settings and screen brightness can help squeeze a bit more life out of your device's battery. Go to **Settings > Display** to adjust them. The **Brightness** setting may also have an auto-brightness option you can check, which dims the screen when indoors. If you use the browser a lot you can squeeze a bit more juice out of your device by going to the browser's **Menu > Settings > Accessibility** page and switching on **Inverted rendering**. This will make text white on a black background, using less power to display pages.

▶ **Install JuiceDefender or an automation app:** JuiceDefender (opposite) helps your device run more power-efficiently. You can also use an app like **Tasker** (see p.122) to automate settings based on where you are, the time of day and other factors.

JuiceDefender is free, but you'll need the **Plus** ($0.99) or **Ultimate** ($4.99) version to unlock some of its more esoteric features.

JuiceDefender Free/$1.99/$4.99 (£1.99/£4.99) 4.6

Squeeze every last bit of juice out of your battery with this automated power manager app. It lets you tinker to your heart's content, specifying precise criteria for (among other things) the behaviour of your device's power-hungry Wi-Fi and mobile data functions, based on criteria such as the time of day, location or current battery level. For example, you can set Wi-Fi so that it's only enabled when the screen is on, but schedule it to open up for a minute every so often to allow synchronization. The included widgets let you easily switch key functions on or off and tells you how much juice you've saved.

Elixir II Free   4.7

System information app with a great
selection of widgets, covering just
about every aspect of your device
from CPU usage to sensor readings
(accelerometer, proximity sensor etc).
Change settings, perform SD and cache
operations, and toggle everything
from USB tethering to screen rotation.

Customizations

Almost every aspect of your phone or tablet's look and feel can
be customized to your needs, from simple tweaks like changing
your icons and widgets, ringtones and wallpapers, to more
complex stuff involving automations and behaviours. Advanced
users can even root their devices (p.125), allowing them to install
customized versions of the Android™ platform.

Google Gesture Search Free   4.5

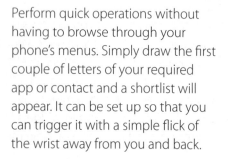

Perform quick operations without
having to browse through your
phone's menus. Simply draw the first
couple of letters of your required
app or contact and a shortlist will
appear. It can be set up so that you
can trigger it with a simple flick of
the wrist away from you and back.

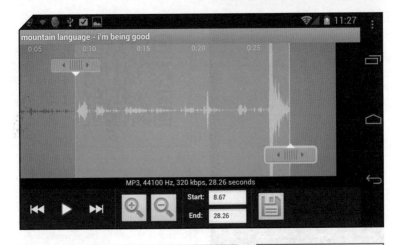

Quick, simple sound editing without any messing around on your computer. If you do want to edit custom alerts outside of your device, save them to the appropriate folder on your SD card (Alarms, Notifications or Ringtones) and they'll be selectable from your device's **Settings > Sound** pages.

Ringdroid Free   4.6

Take complete control of all of your phone's ringtones and notification sounds. Ringdroid lets you pick any song or sound file currently stored on your device's SD card and quickly crop it down to the part you want to use. Edited sounds can be saved as ringtones, alarms, notification sounds or just a straightforward music clip. The interface is intuitive – zoom in and out, swipe your way along the display, tap or press play to listen, and drag the start and end points to taste.

With an almost overwhelming array of functions at its disposal, Tasker is an app for the technically minded tweakmeister in you. Once you figure out how to program it, you'll be able to set up all kinds of automations. It also works with plugins for the similarly versatile **Locale**.

Tasker $6.49 (£3.99)

4.7

A powerhouse of an app that automates functions on your device, based on criteria like its location, the time of day, the phone's orientation, and so on. As a battery-saving trick, for example, you could set Wi-Fi to stop when your screen switches off. Minimize late-night disturbances by setting your phone to automatically go into silent mode between midnight and 8am, but only if your GPS location indicates that you're at home. Make a media player start up as soon as a cable is plugged into the headphone socket, switch your wallpaper from a picture of your cat to something more professional-looking when you arrive at work, and so on. For a free alternative to see if this is your kind of thing, take a quick look at **AutomateIt** from the Market.

Launchers

A new launcher can be a great way to liven up and add new functions to a tired old device. Launchers come in a range of flavours, some, such as **ADW.Launcher** or **Launcher Pro**, offer a familiar take on the home screen experience with a few tweaks and extra features; others, such as **Slidescreen**, attempt novel approaches with varying degrees of success. Others still emulate the user interfaces of rival platforms: **Launcher7**, for example, directs more than a nod and a wink towards Windows 7 phones.

Above and beyond the features offered by Android 4.0's stock launcher, custom launchers often include elements such as a greater number of home screens, animated transitions between windows, extra widgets, scrollable dockers, integrated gestures, shortcuts, downloadable themes and more. They often run faster and smoother than the stock launcher, chewing up less system resources as they go.

ADWLauncher Ex running on an Android tablet.

It's well worth playing around with a few launchers to find one that suits your needs. If you plan on having a few on the go at once, switch between them simply by pressing the **Home** button and selecting which launcher to complete the action with. A check-box at the bottom of the selection panel lets you set your choice as a default. Alternately, grab **Home Switcher** from the Market™, which lets you manage multiple launchers and switch between them easily.

Go Launcher EX Free · 4.6

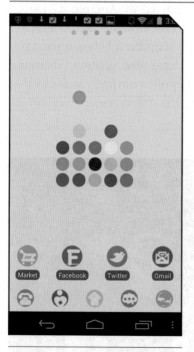

A great place to start if you're thinking of dipping your toe into the wonderful world of launchers. Go Launcher combines the best elements of most of the other heavy hitters, with widgets galore, thousands upon thousands of themes, delightful transition effects, and enough power under the hood to keep you tweaking for weeks.

Go Launcher EX, shown here with a very pleasing free theme from the Android Market called **Color Dot**.

Root access

Android™ may be an open-source platform, but once it's on your phone or tablet, neither the manufacturer, the carrier nor Google™ want you messing with it. Still, there are reasons you might want to "root" or hack your device and unlock access to the core operating system.

Why root your device?

Rooting allows you to install a custom ROM: basically a tweaked version of Android that can add speed, stability and new functionality. For example, the option to install apps to SD memory rather than a device's RAM only appeared officially in Android 2.2, but some custom ROMs already had this feature for quite some time. With the recent war of words and patents between the technology giants, each claiming that the other has stolen their features and ideas, it could be argued that official new releases on the various platforms often borrow more from unofficial custom releases than from each other.

Root access also enables you to run apps that require deeper access to your device's hardware, and allows you to cherry-pick the best custom widgets and user interfaces from rival manufacturers.

Some of the functionality you can achieve by rooting may already be available in your phone or tablet's own brand-specific UI, or through future updates to Android (if your device is still being updated), but it's a crafty way to supercharge an older handset if the manufacturer has been slack in keeping it up to date.

What are the risks?

Well, at best you'll probably void your warranty and waive any further technical support from your carrier. At worst you'll "brick" your handset (it'll stop working altogether, a rare occurrence these days). You'll also lose the ability to get automatic over-the-air updates from the manufacturer.

Even if you root successfully, some performance improvements may mean your phone or tablet's CPU has to work harder, possibly shortening its lifespan or draining more battery power.

So how do I go about it?

Depending on your device and the Android version you're running, the method will vary, but a quick web search including the words "root" and the model of your phone or tablet should yield plenty of results. Tread carefully though, and always read the comments on any blogs or forums you visit to check how successful other users have been.

Obviously, if you do this, you'll be on your own in terms of technical support. If, after considering all the risks, you're certain you want to go ahead, you should always back up all of your data and contacts first (see p.240). A useful primer for the whole process can be found at goo.gl/JP9MD.

Once you've rooted your device you need to test-drive a few custom ROMs. A good place to start would be **CyanogenMod**, a very stable and popular release. For more options, visit TheUnlockr. com and browse to your device for a dizzying array of choices.

▶ **CyanogenMod** cyanogenmod.com

▶ **The Unlockr** goo.gl/dqxpW

For the most up-to-date news on all root-related activity, as well as general tips and tricks, it's worth paying a visit to XDA Developers regularly.

▶ **XDA Developers** xda-developers.com/category/android

mail, sync & productivity

Staying in sync

Email, calendars and more

Now that you have a device that's capable of performing a lot of the same tasks as your computer – email, web browsing, social networking, playing music and video – wouldn't it be great if you could get everything to work across the two devices so that you could move from one to the other without having to manually shuffle files and settings around every time? Well, with Android™ you can do exactly that.

Depending on how you already use your phone, tablet or computer, you'll find that some level of synchronization for email and other online accounts will be up and running at the Android end without much effort on your part. Other things, like your bookmarks and calendars, may be more clunky to sync, but still doable with a little tweaking and some apps from the Market™.

Android integrates seamlessly with Google'™s cloud-based services, so the path of least resistance for keeping stuff synced between your computer and Android devices is to use a web-based Google account (available for free from accounts.google.com, or simply by signing up for a Gmail™ account at gmail.com) as a hub for your mail, calendars and contacts, and then to sync everything with that. There are alternatives but mostly these involve using a bunch of different third-party apps and services instead of Google's. Microsoft Exchange accounts are the exception here; you'll find direct syncing with these, if required, is well supported (see p.131).

Mail

Chances are you have more than one email address, or you have an email address from a provider other than Gmail™. Some people access their email through a desktop client such as Microsoft Outlook or Mozilla Thunderbird, while others prefer to use a web browser to access webmail services. However many accounts you have, and however you normally interact with them, you can keep everything in sync with your Android™ device.

Out of the box, there are two email apps available on your device: **Email**, which can connect directly to your email services in a similar way to a desktop client, and **Gmail**, which is similar but works with a Gmail account. You may think you'd need both, but it's possible to set up Gmail to pull in all your other email accounts so that you can access them all in one place. If you do want to go direct with the email app, though, here's how:

the **Email** app may be all you need for simple but effective account checking.

Email app

To set up a new account with the **Email** app, go to **Settings > Accounts & sync** and tap **Add account** at the bottom of the screen. Here you'll see the various types of account available, based on any apps you have installed that are associated with online services such as Facebook, Evernote, Twitter and so on. You'll also find the option to add an email account.

Syncing with Microsoft Exchange

Setting up your device to sync with your Exchange emails, calendars and contacts is simple. Head to **Settings > Accounts & sync > Add account** and choose **Corporate**. From there just enter your email address and password (if you experience problems, select **Manual setup** and enter your server details). If your older Android™ handset doesn't have this facility already, there are several apps that can sync to an Exchange server, **Touchdown**, **RoadSync** and **ContacsCalendarSync** being solid examples.

From here, fill in your connection details and you'll be ready to send and receive emails. An account's details can be edited later from the email app by opening up its **Settings** page from the Menu button and selecting the account to change.

If your desktop client is set to delete messages from the remote server after downloading them, you may find that you only get emails that have arrived in between your desktop's sync cycles. The only way around this is to set your desktop email client to leave messages on the server, which means occasionally logging in to your email server online to delete messages manually.

Introducing Gmail

 Gmail is Google's webmail service, equivalent to those offered by Yahoo! and Hotmail. You're probably familiar with the concept of webmail, where your emails are kept online (as opposed to being downloaded through an email client and stored on your computer). The main advantages of webmail are that you can access your email from anywhere in the world by simply logging in through a web browser; your mail and contacts are safely backed up and accessible in the event your computer is lost or damaged; and you get stacks of storage space for your emails. The only real disadvantage is that you always need Internet access in order to see them.

> **Tip**: Once you've got a Gmail account, you can use the same login to access other Google™ services like Calendar, Documents, Groups, Picasa™ and many more.

Setting up Gmail on your Android device

When you switched on your phone or tablet for the first time you'll have been prompted to enter your Google account settings, so you may have done this already. If not, head to **Settings > Accounts & sync** and tap the **Add account** button. Alongside other account options, you see one that just says **Google**. Tap this and follow the steps to enter your details. You can set up a new Google account from scratch here, too.

Your Gmail messages (and those from other linked accounts)

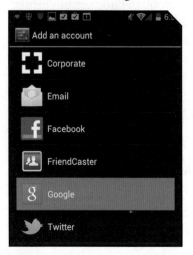

should now be accessible from the Gmail app and any changes you make will be mirrored in your Gmail account online within a few seconds (assuming you have an active Internet connection).

You can add a variety of accounts to sync your device with, including an Exchange server (corporate) or regular email. Gmail is accessible, along with all your other Google-related stuff, from a single Google account login.

A Google account puts all your mail, contacts and calendars in one place. In the course of setting things up to run smoothly with your computer, you'll be visiting the web app a few times and messing with the settings from there.

Using Gmail to access your other email accounts

Gmail can be used to access any other POP3 email accounts you have, giving you a convenient automated backup of all your email and one handy location to access them all from if you're away from your desk.

To set this up, log in to your account with a web browser (your computer's or your Android device's), click on the little Settings cog wheel in the top right corner and select **Mail settings**. Now click the **Accounts and import** tab. About halfway down the page you'll see a section labelled **Check mail using POP3**. Click the **Add a POP3 mail account you own** link and enter the account details in the pop-up window. Gmail will automatically create a label for this account and assign it to any messages that come in through it. You can change this option from a drop-down menu here or leave it for editing later if you need to. If you still want your mail from these accounts to be accessible through the normal routes, make sure to check the **Leave a copy of the retrieved message on the server** box.

Your account will now appear both in the **Check mail using POP3 area**, and also above it, in the **Send mail as...** section.

> **Tip**: At the bottom of the "Send mail as..." section of Gmail's Accounts and Import settings tab, you can select a radio button to reply from the same address to which the message was sent. This will let you reply to emails from any of your email addresses, even from your phone, without switching to another account.

Messages in this account will now be available from the account's label down the left-hand side of your Gmail web page.

From the Gmail app on your device, you'll be able to access any synced third-party email accounts by tapping the Labels button 🖉 from the Action bar and scrolling down to the name of the account (tablet users will see their labels down the left-hand side of the screen and can select from there).

You can use Gmail to add labels to any email based on any

It's worth getting to grips with labels in Gmail. You'll find that like the Groups feature for Contacts (p.138), they give a much-needed extra layer of organization to your messages.

criteria you like, and select which labelled items your Android device bothers to sync. Create new labels and their rules from **Settings > Labels** on the Gmail web page, and select which ones the Gmail app on your device will sync with by going to **Labels > Manage labels**. Tap a label for its sync options: **Sync messages** brings up a panel that lets you sync **None**, **30 days** (the last 30 days of messages will be available to read and search offline) or **All** (all your messages will be downloaded, probably not a great idea for most people). You can also specify custom per-label notifications and ringtones.

K-9 Mail Free   4.3

A popular replacement for Android's stock email app, K-9 Mail is a long-established open-source client supporting IMAP, POP3 and MS Exchange. Features include multi-folder sync, flagging, filing, signatures, bcc, PGP encryption, and the ability to store mail on your SD card.

Contacts

You'll find that while you were busy setting up your Gmail™ account, all your contacts from that account have made their way into your device's contacts list. Fire up the **People** app to take a look.

Now you can create groups, new contacts, delete and edit contacts and perform all other contact-related actions from within your phone or tablet's contacts list, or from your Gmail contacts in a web browser, and the two will sync automatically.

People

People is Android's contact manager. It integrates with Email, Gmail, the Dialler and any social networks you have installed, and can collate and combine all these different sources into one list.

To add a new account, go to **Settings > Accounts & sync > Add account** and select the type you want to add. If you have Facebook, Twitter, Google+™ or other social networking clients installed, those account types will be available here alongside your email accounts.

Enter login details and you're all set. **People** should now show your contacts from those accounts as well.

> **Tip:** At the time of writing, the official Facebook app no longer syncs contacts, although this may well have changed. If it's not working, you can still get your Facebook contacts and their profile pictures into your People list, either by installing **FriendCaster** in addition to, or instead of, Facebook's own app, or by installing **SyncMyPix** from the Market.

After dragging in contacts from multiple sources you'll probably have more than a handful of duplicates. To consolidate two into a single contact, pick one and from its Action bar, select **Menu > Edit** and then **Menu > Join**. From here you'll be able to select the

Syncing Gmail contacts with your computer

Mozilla Thunderbird

Syncing your contacts between Thunderbird and your Gmail account is pretty easy; all you need is the **Google Contacts** add-on (search from Thunderbird's add-on manager: **Tools > Add-ons**). As an alternative, try the **gContactSync** add-on too and see which works best for you.

Microsoft Outlook

Depending on how old your version of Outlook is, there are a number of options available. The best at present seems to be to download and install **GO contact sync MOD**, available from sourceforge.net/projects/googlesyncmod.

Mac OS X

If you're using Apple's Mail and Address Book, you can easily keep your contacts synced with your Gmail account; open **Address Book**, go to **Preferences**, click on the **Accounts** tab, check the **Synchronize with Google** box and then enter your account info. You can then sync the two by clicking **Sync Now** from the Sync menu in the menu bar. Note that you can only sync with a single Google account.

other contacts you'd like to combine it with. If you change your mind, from the same page, hit **Menu > Separate**.

Importing your existing contacts into Gmail

If you're using another webmail service and have a contacts list there that you'd like to import into your Gmail account (and by association, your Android device), all you need to do is log in to gmail.com and click the **Settings** cog wheel (top right of your browser window), then click the **Accounts and Import** tab. You'll see the option to import contacts from Yahoo!, Hotmail, AOL or any other webmail account. From here you can also set up Gmail to pull in email from these accounts.

If you have contacts stored in your computer's email client, Outlook or Thunderbird, say, save or export them to a csv file in DOS format (most email clients can do this). Now log in to gmail.com, press the **Contacts** button and click **Import**. Once imported, these will sync over the air to your Android phone along with your other contacts.

Using Groups to manage your contacts

The annoying thing about Gmail's contacts sync is that it'll automatically pull in every single email address that's in your Gmail contacts list – basically everyone you've ever emailed.

The best way to avoid this is by assigning your contacts to groups, and then choosing which groups your People app (or whatever your device's contact list calls itself) syncs with.

Fire up a web browser and log in to Gmail. Click on **Mail** (top left) and select **Contacts** from the drop-down menu. The first thing you'll then want to do is weed out all the random unknown contacts that you don't need cluttering up your phone or tablet. You don't have to delete these outright, just work through and select any names you don't recognize by putting a tick in the check-box to their left, then click the **Groups** button 👥▾ and create a new group called **Unknown**. Your selected contacts will be

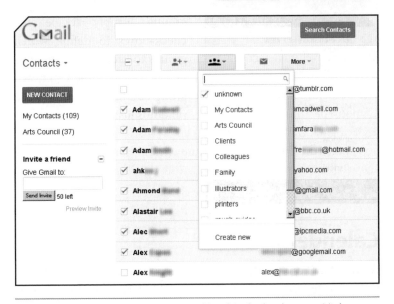

Sorting your contacts into groups can seem like a bit of a drag but once it's done you'll have much more control over whose details appear on your phone or tablet.

automatically added to this group. Now make sure that the **My Contacts** group in the drop-down list is unchecked for these contacts. You now have two groups: My Contacts (the default group), which contains people you know, and Unknown, which contains people you don't. You can of course create as many groups as you like, and contacts can be assigned to multiple groups.

Back on your Android device, fire up **People**, and from the contact list tab select **Menu > Contacts to display.** Here you'll see a list of any accounts you have synced for contacts (including any social networks you're syncing with, see p.136). Select **Customise…** from the bottom of the list and tap the entry for your Gmail account to expand it. You'll see a list of all your contact

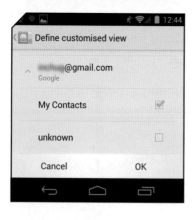

groups. Simply toggle the check-boxes to define which groups to keep synced on your device. Uncheck the Unknown group and hey presto, all those randoms just disappeared.

Calendars

As with mail and contacts, the easiest way to get your Android™ Calendar to sync with any other calendars you already have is to move those calendars to your Google™ account and then sync everything to that. Once you have this set up, your Google Calendar™ will sync automatically with your Android device and your computer over Wi-Fi.

To move a calendar to your Google account, you first need to export it from its current location into iCal (Mac) or csv (MS Outlook) format. Outlook users, you'll need to fire up your calendar and then go to **File > Import/Export**. Thunderbird users, go to **Events and Tasks > Export**. ICal users should select **File > Export**. Select the calendar to export, the appropriate format (iCal or CSV), and the location (browse to somewhere you know you'll be able to find it, like your documents folder or desktop).

Next, open up your web browser, log into your Google account and hover your mouse over the Google logo (top left), select **Calendar** from the drop-down list that appears (you may need to click **More...** to see the Calendar option). Once on the Calendar page, click the cog-wheel (top right) and select **Settings**, and then the **Calendars** tab. Now click **Import Calendar** and browse to the exported calendar you saved earlier.

Your calendar is now a Google calendar. It's probably synced with your Android device already. Now all we need to do is get it syncing with your computer (see box on p.142) and you're ready to rumble.

The Calendar app

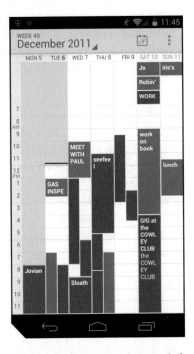

The Calendar app will sync with as many calendars as you can throw at it. You can manage which calendars you're syncing with from **Menu > Calendars to be displayed > Calendars to be synced**. Android assigns a different colour to each calendar and displays them in the standard variety of formats – by day, week, month, or as an agenda – selectable by tapping the date at the top, or from the options above that for tablets. You can zoom in and out of the various views by pinching and un-pinching with two fingers, and swipe left and right for the previous or next day or month. Tap an empty space to create a new event, or tap an existing event to see it in more detail, add reminders, or to edit it (the ✎ button in the Action bar).

Syncing Google Calendar with your desktop client

Microsoft Outlook

Most users will find **Google Calendar Sync** (tinyurl.com/22yq2m) perfectly adequate for their needs. Once downloaded and installed on your computer, simply enter your Google account login details and select **2-Way sync**.

If you need more robust capabilities, including full contacts, calendar and mail syncing, it may be worth investigating a subscription to Google Apps Premier Edition, which allows you to use the **Google Apps Sync for Microsoft Outlook** application. Visit goo.gl/mlmZS for more details.

Mozilla Thunderbird or Sunbird

Mozilla users can download the **Provider for Google calendar** add-on. This allows you to view and edit your Google calendars directly from Sunbird, or Thunderbird's **Lightning** calendar. Once installed, right-click in the calendar area and choose **new calendar**. Select **On the network** as the calendar's location, and on the next screen, **Google Calendar** as the type. In the **Location** field you'll have to paste in your calendar's full address, which you can find in the settings area of your Google calendar by clicking the green **ICAL** button.

iCal

Mac users can sync iCal calendars with Google Calendar by going to **Preferences > Accounts** in iCal and entering your Google account info. You'll need to then go to **Delegation** and select the particular calendars you want iCal to access. Once you're in sync with Google, you'll be syncing with your phone.

Not running the current Mac OS? Check out the instructions for syncing earlier versions: tinyurl.com/56byzz.

Jorte Free

Pleasant-looking and easy to use personal organizer-style calendar replacement. Has most of the functionality of the stock app with added tasks, schedules, and more. It may take a little tweaking to get it syncing with your Google calendar, but it's well worth the effort.

Quick Event Free

Quick Event lets you quickly add events by voice entry. It supports recurring events, date-spanning and other complex details. You can say something like "On location from 2pm tomorrow until 5am Friday, in London," and the event will be created with times, dates and location all present and correct.

Touchdown HD Free

Comprehensive Outlook-like app for MS Exchange accounts, bringing email, contacts, calendars and tasks, notes and SMS together into a single tabbed interface. It's fully customizable with an impressive array of widgets. An excellent all-in-one solution for corporate users.

Productivity

Everything else you need to stay busy

If you plan on using your Android™ device as an extension of your office, you'll need more than just email and calendars. Fortunately the Android Market™ is brimming over with document editors, to-do list managers, note-takers and other apps to help you stay productive while on the go.

Office suites

Google Docs™ Free   4.1

Google's own tablet-optimized Docs app hooks you up instantly with your Google Docs online for viewing and editing spreadsheets, presentations, word and text files, images and more. For a free suite it's excellent for lightweight word processing needs. Anything more complicated (editing spreadsheets, for example) and you'll probably want to look to one of the paid apps.

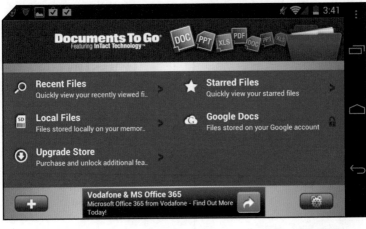

You wouldn't use it to write your memoirs but it's handy for last-minute editing while on the move.

Don't let the high price tag put you off – Documents To Go is frequently on offer in the Android Market, Amazon Appstore and other places around the web.

QuickOffice Pro ($14.99) is another highly regarded office suite worth checking out.

Documents To Go $14.99 (£9.99) 4.3

Lets you view, edit and create Microsoft Word (.doc and .docx), Excel (.xls and .xlsx), PowerPoint (.ppt) and Adobe PDF files (its PDF viewer is one of the best available). Impressively sophisticated for its size, it offers features like word count, find on page, formatting tools, word wrap, multiple zoom levels and track changes, and support for password-protected files. There's also a Live Folder facility that puts a recently used files folder on your home screen. It also syncs and edits Google Docs files with ease.

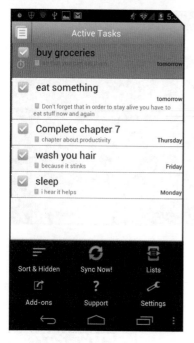

To-do lists

Astrid Tasks Free · 4.4

To-do lists aren't exactly the most exciting thing on the planet, although it does depend somewhat on what you've got on your list. If yours is full of all the nasty stuff that you just wouldn't get on with unless there was either a really big carrot or a really big stick in the vicinity, you're not alone. Astrid is an encouraging little pink octopus thing that arranges your to-dos and sends you quirky little reminders, usually more carrot than stick: "A little snack after you finish this?", or "Ready to put this behind you?". Behind the cute facade, Astrid is a GTD (Getting Things Done™, as pioneered by David Allen) powerhouse, web syncing with Google Tasks™, Producteev and Astrid.com. It supports multiple lists and list sharing, and a handful of paid-for plugins, allowing you to add tasks via voice, trigger reminders based on location and more. If Astrid's friendly interface just isn't your scene, **Wunderlist** is another great task manager worth a look.

Notes and ideas

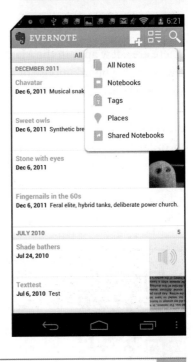

Evernote integrates nicely with your phone's share functions. Hit **Share** or **Send** from your camera and most other apps and you'll find the option to send the file you're working with to Evernote without even having to start up the app. For more storage and organization of your notes, bits and pieces, check out the equally brilliant **Springpad**.

Evernote Free

4.6

Evernote looks like a humble note-taking service, but beneath its surface lies a powerful web-based capture tool. Register for a free account at evernote.com and the app will let you send text and audio notes to yourself, upload photos, scanned text and other files, tag them and access them from anywhere via your web browser. Any writing in images you upload gets transcribed into searchable text. It's a neat way to keep all your ideas, reminders and notes in one place. It integrates with free desktop apps for PC and Mac and there's also a paid account option offering more storage space.

Remote control

There are a few apps in the Market™ that let you use your device as a wireless keyboard, mouse or touchpad to control your computer. **RemoteDroid** is a highly regarded free wireless touchpad app, as is **GRemotePro**. These kinds of apps usually require you to download and install a server application to your computer in order to work, so there may be a small amount of setting up involved, but nothing too complicated. For more complete control of your computer, try **PhoneMyPC**...

PhoneMyPC $14.99 (£9.29)   4.5

One of the many apps on the market that lets you control a computer from your phone or tablet over Wi-Fi or a mobile data connection. Install the app on your device and the software on your PC, set up a password and you'll have a secure link between the two. You can operate your computer more or less as if you were sitting in front of it, using the phone's touch screen, or just take a quick snapshot to check the progress of a download. You could open up a program, make some changes, save them and email the new file to yourself – all from the other side of the world. Assuming, of course, that you remembered to leave your computer on. There are quite a few apps on the Market capable of this kind of thing, a search for VNC (Virtual Network Computing) will serve up a whole range of options. Other popular apps worth considering include **Remote Desktop Client**, **LogMeIn Ignition** and **PocketCloud**.

Money

Invoice2Go Free/$9.99 (£6.99) 4.4

Make fast, professional-looking invoices, credit memos and purchase orders from customizable templates and email them as PDFs (including optional PayPal buttons for instant payment). Invoice2Go calculates totals and taxes based on your location and currency and helps you keep track of what you're owed. The free version restricts you to storing three invoices at a time, any more than that and you'll need to delete them or buy the full app for $9.99. As an alternative, try **Zoho Invoice**, which is free for up to five invoices per month, with monthly fees kicking in above that.

PayPal Free 4.2

The Android™ PayPal app provides some simple tools for securely sending and receiving payments from your PayPal account. Aside from general account management, the app offers a few other features, such as a bill-splitting calculator and the ability to set up payments over NFC by bumping two (NFC-enabled) devices together. US users can find local businesses that accept PayPal payments, and cash a cheque into their PayPal account simply by taking a photo of it.

Need a one-stop snapshot of your finances? Planning a budget? Look no further than Pageonce. For a similar service, take a look at **Mint.com Personal Finance**.

Pageonce Free   4.5

Pageonce is a secure web service that lets you consolidate all your financial information – bank accounts, credit cards, bills and investments. Track and pay all your bills – with due date alerts and email reminders – monitor your transactions and view detailed statements. Track frequent flyer miles and rewards, and your phone minutes, text and data usage. You can also view detailed reports and charts analysing where your money's going. If you have accounts strewn across the Internet, you can wave goodbye to constantly logging in and out of them all and manage everything from one place.

Creativity

Adobe Touch suite $9.99 (£6.99) each

4.0

Adobe, makers of the ubiquitous Photoshop, have entered the Android Market™ with a suite of tablet-oriented touch-screen apps aimed at creative types. The apps are $9.99 each and include **Collage**, for assembling mood boards combining drawing, images and text; **Kuler**, for generating harmonious colour themes; **Ideas**, a vector-based digital sketchpad (below); **Proto**, for building wireframe prototype website designs; and **Photoshop Touch**, which ports core functions from the desktop photo-editing program to a more intuitive touch interface. Powerful stuff, but before you dive in, you may find **Fresco Pro** or Autodesk's **Sketchbook Pro** provide what you need for less outlay.

Other productivity apps

Gleeo Time Tracker Free 4.7

Simple time recording tool that lets you break down a project into as many component activities as you like and separately time-track each. You can easily track multiple projects and access statistics over an animated timeline.

Parcels Free 4.7

Allows you to track parcels from all the major shipping services in the US and UK. Track multiple packages, labelled and colour-coded for easy reference, and receive detailed status updates. It can even display progress on a Google Map. Want more? Check out **TrackChecker**.

ezPDF reader $2.99 (£1.90) 4.6

There are plenty of free PDF readers on offer, including Adobe's own **Reader** app, but if you're after something a bit more solid, ezPDF is worth the cost. It displays accurately, has sensible small-screen enhancements and lets you add annotations and complete forms.

Internet

Using Android to get online

You've probably cottoned on by now that your Android™ device without an Internet connection is like a hamster without a ball; it just wants to run around exploring stuff and bumping into things but instead it's stuck wandering around the small, familiar environment of its cage, absent-mindedly seeing how many sunflower seeds it can hide in its face.

Most of the apps and features covered in this book require at least semi-regular Internet access in order to function properly, the most obvious culprit being your device's web browser. With all of the apps available that are designed to provide their own interface for what are essentially websites (social networking apps, Gmail™, eBay, and so on), you may not be using your Android browser nearly as much as you'd use the browser on your desktop. It's easy to forget that many of the apps and services you take for granted also have mobile-optimized versions of their websites that you can access through the browser. Most sites will auto-detect your browser type and automatically redirect you to the mobile interface; a simplified arrangement of the site's key features that often resembles and occasionally surpasses the site's own dedicated app. If you're not automatically redirected to a site's mobile version, you can usually find it by browsing to m.sitename. com, mobile.sitename.com, touch.sitename.com, or sitename.com/m. If you do ever want to direct your browser to the normal version of a site (to use features not available from the mobile site, for

Same site, different look: Facebook's mobile site (right) offers most of the same features as the desktop site (left), but lets you access it in a manner more appropriate to the small screen. The mobile interface is largely indistinguishable from Facebook's native Android app, which may make you wonder why you'd bother with the app at all.

example), you can get to it by tapping the browser's **Menu** button and selecting **Request desktop site**.

The browser

The stock Android™ browser is based on the same open-source WebKit engine as used by Google's Chrome™ browser. It's pretty snappy for general use, and comes armed with some handy features.

On tablets, most of the browser's features such as bookmarks and tabs are in plain sight. Phone users may need to dig a little deeper into their menus to access these items.

Starting up the browser, you'll be taken to the last web page you were looking at (or, if this is your first time, you'll be looking at Google.com). Drag the page down slightly to reveal the address bar, tabs and menu button (tablet users, all this stuff will probably already be visible to you). From here you pretty much use it like a normal web browser: tap in the address bar to bring up the keyboard and enter an address or search term, then tap the **Go** button (bottom-right corner of the keyboard).

Tabbed browsing

Tabs let you have more than one site visibly on the go at a time and jump easily between them. On a tablet, you'll see the tabs running along the top of the screen; to create a new tab, just tap the **+** in the tab bar. Hit the **X** on any tab to close it. On a phone, you can

Tabs are a little more fiddly on a phone, but still only a few taps away.

access tabs by dragging down to reveal the address bar and tapping the ▤ button. Here, you can quickly switch tabs, or swipe sideways to discard them. Create a new tab by tapping **+** in the Action bar. You can also access your bookmarks from here by tapping ★.

To open any link from the browser in a new tab, long-press it and the option will appear in the pop-up menu.

History, bookmarks and saved pages

To bookmark a page, press the ★ button (tablets) or select **Menu > Save to bookmarks** (phones). You can also select **Menu > Save for offline reading** to keep a page handy for when you don't have Internet access.

Long-press your device's **Back** button to access the browser's History pages. The tabs at the top of the screen give you access to your bookmarked pages and pages you've saved for offline viewing.

Bookmark syncing

The Android browser is able to automatically sync bookmarks with Google's **Chrome** web browser. If your Chrome bookmarks haven't

The stock browser plays nicely with Google Chrome, if you have it installed on your computer. Sync bookmarks, passwords and more.

shown up in the browser already, open up Chrome on your computer and click **Menu** (the spanner button, top right) > **Set up sync**. You'll be prompted to sign in to your Google account and select which information you'd like to sync (bookmarks, saved passwords, and so on).

Back on your Android device, open up your bookmarks and you should see, in addition to your local bookmarks, an extra bookmark list named after your Google account. If this hasn't appeared, go to **Settings > Accounts & sync**, tap on your Google account, and make sure **Sync Browser** has a tick next to it.

To sync from other desktop browsers (Firefox, Opera, etc) your best bet is to install the Android version of that browser and sync with that, or to use a third-party solution such as xmarks.com and use the associated app from the Market to sync things up at the Android end.

Incognito mode

Incognito mode lets you browse the Internet with the relative privacy of not picking up any cookies or generating any web or search history. You can open a **New incognito tab** from the **Menu**

The end of Flash?

Adobe have decided it's time to phase out support for Flash on new mobile devices. The final release does support Android 4.0 devices but future Android updates will likely be Flash-free. Even as recently as a year ago, this would have been a major stumbling block for accessing many websites, but lots of the tricks Flash kept up its massive, security-hole-riddled sleeves are now achievable in HTML5, which most modern browsers support. The upshot of this is that for the most part you won't notice any difference when browsing the web without Flash installed. You may find some sites, such as the BBC's iPlayer, no longer work with your device, but you can get around this specific problem by installing **myPlayer** from the Market, or for Flash video content generally by using the **Skyfire** browser (see below).

button on tablets; phone users will find the option hiding in the menu only when viewed from the **Tab** window. A new tab will open with a secret agent icon in the top left.

Skyfire Free/$2.99 (£1.79)   4.3

Excellent browser with some powerful features. Enables you to watch Flash video content online by transcoding it into HTML5 on Skyfire's servers (this feature is free for three days, after which you need to pay $2.99 to unlock it). It also has some unique social networking integration, with built-in "like" buttons, a feed reader, and "related ideas'" search.

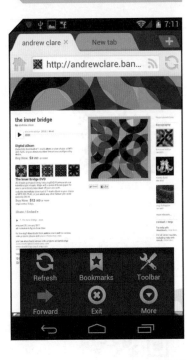

Dolphin in action. Drag to the left for a hidden toolbar, drag to the right for all your bookmarks, which can then be dragged and rearranged. The tabs and address bar can be hidden by entering full-screen mode (available from the sidebar, so you don't have to exit out to a settings page to get them back).

Dolphin Browser HD Free   4.6

A lightning fast, heavily featured web browser. Something of an Android veteran (it's now up to version 7), Dolphin just seems to get better and better. It supports add-ons (more than sixty currently available) and themes, multi-touch pinch zoom, tabbed browsing, RSS feeds, speed-dial, bookmark sorting and sync with Google™ Bookmarks, a customizable menu bar and full-screen mode. It also sports a password manager and assignable gesture commands. All this, and it's free. If you're looking for an upgrade from the stock browser, Dolphin (or its snappier, slimmed-down cousin Dolphin Mini) should be your first port of call.

Opera Mini Free   4.4

Great little browser that compresses web sites before they reach you, meaning you can load pages faster and using less data. Ideal for browsing on a limited data plan. It also syncs bookmarks with Opera on your desktop computer. For more features check out **Opera Mobile**.

Firefox Free   3.5

If you use Firefox on your computer and need bookmark, tabs, history and password sync or excellent add-on support, this could be your answer. Tablet users will also appreciate its new tablet-optimized layout. Firefox can be a bit of a heavy, lumbering beast for a mobile browser, so unless you need Firefox sync you may be better off with Dolphin.

Miren Browser Free   4.4

Very fast, lightweight browser with some convenient features, such as swiping left and right to go back and forwards, page scrolling with the device's volume buttons, brightness controls, tabs, bookmark folders, and plenty more.

Social networking

Social networking is a doddle to set up on an Android™ phone or tablet. Depending on your device, it may have shipped with **Facebook**, **Twitter** and **Google+**™ apps built in; it may also have prompted you to set these accounts up when you first switched on; and it may even have combined Facebook or Google+ photos and contact details with people already in your contact list (or not, see p.136).

The official apps for these and other social networking sites are available for free from the Market™, along with a dizzying array of third-party equivalents. It's worth starting with the official apps and then trying some of the others if they offer extra features you feel you're missing. Once installed, you can set how often you'd like the app to sync with your online account, and which kinds of data you'd like to see. You'll usually find detailed settings hiding in menus within the app itself, with more general sync settings available from your main **Settings > Accounts & sync** screen.

> **Tip:** To integrate contacts from your social networks into your main contacts list, sync photos and manage multiple contacts, see p.136.

Path Free   3.0

An interesting new mobile social network, Path takes a journal-like approach, allowing you to share thoughts, photos, videos, music and events with up to 150 friends. The Android app currently falls short of the iOS version, but, hopefully, this will improve over time.

The official Facebook app seems to change every other week: features appear and disappear, the interface reshuffles and reinvents itself with every update. At present it's indistinguishable from the mobile site, which is actually quite a good thing. General consensus, though, seems to be that the best alternative out there is Friendcaster.

Friendcaster/Pro Free/$4.99 (£3.04)   4.1

Friendcaster manages to squeeze in most of the missing functionality from the official Facebook app. It has reliable notification support with a quick-reply facility, pull-to-refresh status updates, add/remove and favourite lists for friends, in-app photo editing and tagging, easy link sharing from the browser, events and groups, built-in chat, plus some nice-looking widgets for grabbing a quick update from your home screen. It also lets you select privacy settings for individual posts and supports SSL encryption for added security. The app is free, with an ad-free version also available for five dollars.

Plume arranges tweets, messages and mentions in different columns you can swipe between (tablet users will see all three columns arranged horizontally on the same screen). A fourth column lets you save custom searches.

Plume/Plume Premium Free/$2.99 (£1.79) 4.3

Nice-looking Twitter client with a customizable interface. It supports multiple accounts, geotagging, URL shortening and Twitpic photo upload. It auto-completes hash tags and user names, and lets you colourize or mute individual tweeters. Drag the bar at the top for a pull-down area for adding your own tweets. It displays inline conversations, profiles and replies, and has some beautiful widgets (as you'd expect from the makers of a collection called **Beautiful Widgets**). Plume is free, or you can fork out $2.99 to make the ads disappear. Other fine Twitter clients include **Tweetcaster** and **Twicca**.

Google+™ combines elements of Twitter and Facebook and adds a few tricks of its own. It also integrates neatly with other Google services you may already be using (Picasa™, Reader™, Gmail™ and others). An interesting addition to the social networking realm, if only more people were using it! Certainly one worth keeping an eye on.

Google+ Free   4.3

Google's attempt at a social network is still in its infancy, with the mobile app providing most of the features supported by the site. The Stream section gives you fairly standard issue timeline of the people you're following, allowing you to "+1" an entry (Google's equivalent of a "like"), add comments, re-share and so on. Other sections let you view and tag photos, manage your circles and the people in them, and group chat. It syncs contacts seamlessly with your main contact list (you can choose which circles to sync) and has a neat instant upload option, which sends any photos you take directly to your account.

Combining your social networks

Wonderful as it is being able to access all these different streams of information coming in from all over the web, it's a pain to be constantly jumping between web pages and apps. Enter the snappily named social networking aggregators, a group of apps that pull everything together in one place. Your device may well have shipped with one – Sony Ericsson's is called **Timescape**, while Motorola has **Motoblur** and HTC offers **Friend stream**. Any of these will draw in news feeds, status updates and other information from your synced accounts and give them to you in one long list. If you don't have one already, or are looking for something a little more versatile than the one that's been foisted on you, here are a few worth your consideration.

Seesmic Free 4.3

Seesmic is a decent, simple aggregator that hooks you up with your Twitter, Facebook, Google Buzz™ (does that even still exist? Hopefully Google+ integration is just around the corner) and Salesforce Chatter accounts. Facebook integration is fairly basic; it lets you read and post status updates, comments and likes, and manage any pages you administrate, but stops short of providing Chat, photo albums, or many of the other features offered by a dedicated client. The Twitter side of things is better equipped, supporting multiple accounts, cross-posting, lists, retweets, conversations and more.

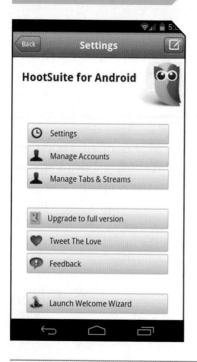

HootSuite has the edge over some other aggregators in that it also provides a high-quality web client, meaning you can log in from your desktop and still get all your networks in one place using the same client.

HootSuite Free

4.2

Aimed more at PR professionals but equally just as by the average Josephine, HootSuite combines Twitter, Facebook, Foursquare and LinkedIn streams into a single screen for each identity you have, keeping track of mentions, messages, retweets and so on. You can post updates to all your networks at once or pick individual ones, and syndicate RSS feeds from your blogs or anywhere else into your accounts. Basic accounts are free, and include enough functionality for most non-corporate users.

photos
& video

Photos & video

Your device's camera and more

Whether your Android™ device is a phone or a tablet, it will almost certainly have a camera built in. It may even have two: a main, back-facing camera for taking pictures and video, and a lower-quality front-facing camera for video chat (p.91), face unlocking (p.238), or for just checking your hair.

Even the best phone cameras are inevitably a poor match for a standalone digital camera because the optics are so tiny, but the sensors behind those optics are becoming incredibly refined – your device may be capable of shooting up to 1080-pixel HD quality video – and Ice Cream Sandwich comes equipped to take full advantage of all that sophisticated hardware.

The Camera app

 Probably the first thing you're going to want to do with the built-in camera is to take a few photos. Your device may have shipped with a brand-specific camera app, or it may have the stock Android camera (as described below).

Whichever you have, it's possible to swap this out for something a little more capable from the Android Market™. Camera replacements are available by the bucketload (p.175).

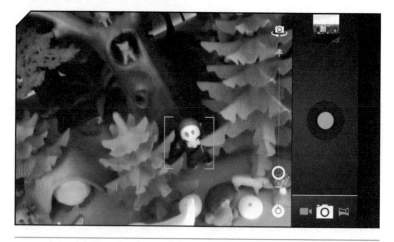

Camera mode. Your navigation buttons are still there to the right of the screen but have quietened down into dark grey dots in order to be less of a distraction.

Taking photos

Firing up the Camera app will take you by default into standard photo mode (as opposed to panoramic or video modes, which are selectable from the buttons at the bottom-right of the screen). To take a shot, simply tap the big blue circle to the right of the display.

You can set the point of focus and exposure by tapping the area on the screen; a square bracket will momentarily appear to show you where your focus and light reading is targeted (as in the image above) and the display will adjust accordingly.

To the right of the image, you'll see a zoom bar. The 🔲 button above this switches between your device's front- and back-facing cameras, while the ⏻ button at the bottom opens advanced settings (flash, white balance, exposure time, scene mode, picture quality and location tagging).

Video mode operates in much the same manner as Camera mode. Settings and the ability to switch cameras are switched off while recording, but zoom is still available.

Shooting video

Shooting in video mode is as simple as tapping the big red button to start filming and tapping it again to stop. The camera will automatically adjust light and focus as you film, and there's no option to change this manually.

Tip: You can tap the image at any point to take a photograph without ending the recording.

Settings and camera switching become available between takes, above and below the zoom bar. As well as standard adjustments like white balance and image resolution, the video settings offer a few interesting effects to play with. Tap ⤵ for time-lapse mode, or ▨ to choose from horrific "silly faces" effects, giving you hideously distorted eyes or an enormous gaping mouth with which to astound and terrify your loved ones.

Here's one I made earlier: it's actually a little bit scary just how easy it is to take a panoramic shot like this.

Panoramic photos

To take a panoramic shot, select the ▤ icon from the camera mode button. Tap the big green button to begin and slowly rotate or pan the device across your chosen scene. Once you're done, the camera will chew on the data for a while as it knits the panorama together. That's pretty much all there is to it. If your final image looks a bit garbled, try taking the shot again a bit more slowly. If panoramic shots aren't automatically available on your device, try **Pano** or **Photoaf Panorama** from the Market.

HDR Camera+ $3.99 (£2.53)   4.3

Takes amazing images using HDR (High Dynamic Range), a photographic technique where the same scene is shot several times over a range of exposures, which are then blended together to create a more intense, detailed image.

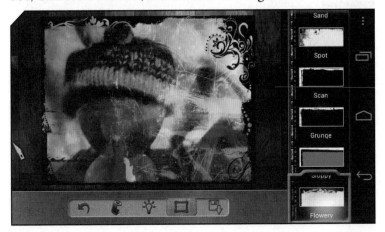

Pixlr-o-matic Free 4.6

Offers a beautiful array of adjustments, filters, retro effects and custom frames to give your photos some added pizzazz. There are plenty of other photo effect suites out there to choose from, some excellent examples being **Camera 360**, **Camera Zoom FX**, **Little Photo** and **Vignette**.

After testing a range of movie editing apps, it seems that Android's own Movie Studio is the best one out there. If you're unable to find it for your device, try **AndroMedia HD** for a similar set of features.

Movie Studio

 Android™ 4.0 ships with this useful movie editing app. You can import your captured video and images, trim and splice them together, add effects, titles, captions and transitions, and then export your finished film to a range of formats.

To get started in Movie Studio, tap Create New Project or the ▣ button (top right) and name your project. Your new project screen consists of a preview area at the top and a timeline at the bottom. Tap ▣ to add photos or video clips to the timeline (either from existing footage or shot directly from the app), and tap the + either side of it to add additional clips before or after. Clips display as a series of still frames showing what's happening in the footage relative to the timeline. You can compress or expand the timeline display by pinch-zooming in or out.

Editing clips

Tap an individual clip to edit it, and a blue outline will appear around it. Drag the blue triangular handles left and right to trim the clip. Straight down the centre of the timeline you'll see a vertical blue keyline. If you think of your timeline as physical video tape, you can think of this blue keyline as the playhead displaying that point on the tape in the preview screen above. Dragging the

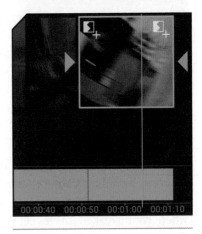

Close-up of the timeline, showing an edit in progress.

timeline left and right over the playhead changes the preview accordingly. Similarly, when pressing Play in the preview, you'll notice the timeline scroll along in time with the footage.

Tap the ▣ in either of the top corners of the clip to add a transition effect (cross-fades, wipes, etc) at that end of it. While a clip is selected, you can add effects and titles to it from the Action bar.

To add sound or music, tap the ◀ at the bottom of the screen and select your track. A blocky orange waveform appears to represent the sound source. You won't be able to do much sound editing other than adjusting volume or ducking (dropping the volume of your music track while someone in your film is talking).

When you're happy with your movie, you can save it by selecting **Export Film** from the menu. Once exported (it may take a while to render your project into a new movie file) you can select **Share film** from the menu to send it to a variety of sources, including your YouTube™ account if you have one.

The Gallery app

 The Gallery brings together all your local photos, screen grabs and videos, along with any images in your linked accounts (Picasa™, Google+™, Friendcaster for Facebook, and so on). As well as providing a single location to access all your visual media, it also provides some inbuilt editing tools and special effects.

The Gallery opens as a grid of folders for different file types and locations, including **Camera** (any photos or videos you've taken); **Movies** (edited movies exported from Movie Studio); and **Edited** (where your tweaked images get saved). Other folders house synced images with your linked accounts online (Google+ profile pictures, Facebook photo albums and so on). Long-press **Albums** in the top left to select other views, sorting your images by location tags, time-stamps, or any other tags they may have.

In any of the grid views you can select an item by long-pressing it, and then select as many subsequent items as you like by tapping on them. While in selection mode, the Action bar changes to present sharing and deletion options.

Swipe through the grid of pictures and tap on one to open it, then you can swipe left and right to view previous and next shots. Double-tap or pinch to zoom in and out.

When viewing single images, tap once to bring up the Action bar and scrollable thumbnails of the album's contents. From this view you can delete the image on screen, share it from the drop-down share list, or tap the menu button for a list of other options, including a slide show, and edit, rotate and cropping facilities.

Adding effects

When viewing an image, select **Edit** from the menu to make adjustments and add effects. A toolbar at the bottom of the screen allows you to adjust light and shadows, add some cool vintage effects, make colour adjustments or doodle on top of the image, as well as the more mundane rotate, flip, straighten crop and other useful photo edits. Undo, redo and save buttons are at the top of the screen, while tapping your device's Back button exits this mode completely without saving.

Goodbye Photoshop; some beautiful photo filters are available directly from the Gallery.

179

Screen grabs

New in Android™ 4.0 is the facility to make a quick and easy screen grab. Simply press the **Power** and **Volume down** buttons at the same time. Hold them down for about a second and you'll see a snapshot being taken. Screen grabs can be viewed later from their own sub-section in the Gallery. Grabs are stored on your SD card as hi-res .png image files, and can be found in the **Pictures > Screenshots** folder (not in the DCIM folder where your photos are usually stored).

Syncing your photos

There are a number of ways you can sync photos with your computer, the most basic being to connect the two devices via USB and use the computer to browse to the DCIM folder on your device's SD storage. You can also use **DoubleTwist** and **AirSync** (p.194) to wirelessly transfer photos between your devices.

If you have Google™'s **Picasa** installed on your computer (available for free from picasa.google.com), any folders for which you have **Sync to web** switched on will automatically upload to your Google+ albums and sync with the **Gallery** on your Android device. At the time of writing, this doesn't work for HTC phones or tablets running the Sense UI. In this instance, your best bet is to sync with Flickr using the apps supplied with your device.

Tip: To set your device to automatically upload any photos or video you shoot to your Google+ albums, from the **Google**+ app, select **Menu > Settings > Turn on Instant Upload**.

media & games

Movies & music

All your entertainment needs

Google™ seems to have an ongoing mission to provide a complete iTunes-like experience for Android™ users. This is mostly unfolding through the Market™ (both the website and Android app) along with integrated Google-branded apps which interact closely with each new section of the Market as it emerges. As well as installing apps and games, you can now use the Android Market to download books (see p.200), rent movies and (only in the US at the time of writing, although likely to follow internationally) buy music.

As with apps and games, you'll need a Google Checkout™ account and valid credit or debit card in order to make use of these services. If you don't already have one it's easy to set up and you can do it at the same time as making your first rental or purchase (see p.71–72 for setting up Google Checkout).

Movies

Your Android™ phone or tablet is a powerful multimedia device capable of playing back high quality HD video. Movies in a variety of file formats can be synced to your device from your computer, or streamed from the Internet by renting from the Market™, as well as other sources (most of the major rental outlets now provide Android apps). There are some excellent media player apps available to install, and, of course, access to services like YouTube™.

Videos Free   3.3

Google's Videos app, available from the Market, is a simple, easy to use movie player. It consists of two tabbed areas: **My Rentals**, which gives you access to any films you've rented from the Android Market (see below), and **Personal Videos**, which collates any footage captured with your device's camera and any other video files kicking around on your device. You may have a preferred media player already installed (see p.188–189 for some good examples), but if you intend to rent movies from the Market, you'll need to have Videos installed.

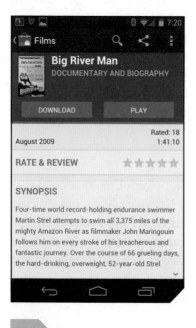

Renting movies from the Android Market

Movies can be rented by firing up the Market and tapping the **Movies** tab (UK users will see a tab called **Films**). Alternatively, from the **Videos** app, tap the Market icon in the top right to go directly to the Market's Movies section. Peruse films by category or use the search bar to find something specific.

The Market gives you the option to play a stream of a rented movie, or download it for viewing later (see p.185).

The rented movie, ready to watch in the **Videos** app. Once you hit Play, you've got a day (or sometimes two) to watch it.

Once you find something interesting, you can get more information and see trailers by tapping on the picture. Tap the **Rent** button and follow instructions from there. You'll usually have a thirty-day window to start watching your film (although this may vary depending on the distributor). Once you start though, usually you'll have to finish watching it within 24 hours (although you can watch it as many times as you like within this time frame). Movies will play back on your Android device, computer or TV (assuming you can connect it to either of the above with an HDMI cable) and even through a linked account at YouTube (with added movie extras). For more details about watching rented movies on other devices, check the guide here: http://goo.gl/ItsLR.

Pinning movies for offline viewing

If you'd like to save a movie to your device for watching later – while travelling, for example – you can do this by "pinning" the movie to your device. From the **Videos** app, tap the Menu button and select **Manage Offline**. You can then select movies to pin by tapping the 🔄 button next to each one you'd like to download.

Movie refunds

If you decide not to watch a movie after all, you can cancel it within 7 days for a refund (so long as you haven't pressed play). You can't actually do this from the Market or the Videos apps though; to get a refund you'll have to head to the Android Market website, select **My Library** from the top of the page, then click **My Orders** (top right) and click **Report a problem** to the right of the movie you'd like to cancel. Fill out the form that pops up and be sure to put a tick in the box at the bottom that says you'd like a refund. Alternatively, you can log into Gmail™ and find the email receipt notification relating to your movie rental, scroll to the bottom of the message and follow the links and cancellation instructions from there. It's a surprisingly convoluted process but so long as you haven't started playing the movie, your refund should go through without any problems.

Pinning a film so you can watch it on the train home, with a complete stranger's lolling head gently depositing a string of drool on your shoulder.

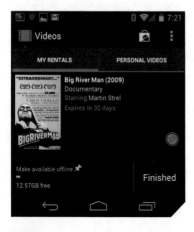

Once you've made your selection, tap **Finished** and your movie(s) will start to creep down the line. If you like, you can start watching the movie while it's still downloading, but it can take an hour or more to completely download a film, so make sure you leave enough time if grabbing something to watch before setting off on a journey.

Tip: Pinning a movie for offline viewing will disable streaming of the movie on your computer or other devices. Unpinning will re-enable it for streaming or downloading onto additional devices.

Other ways to rent movies

If you already subscribe to a movie download service like Netflix or Hulu, you'll find apps for these on the Market that let you use your Android device to stream films in much the same way as you'd use your computer or set-top box. Some services, like the UK's (Amazon-owned) LoveFilm haven't quite caught up yet but it's only a matter of time.

Streaming live TV

There are hundreds of individual TV channels (or sometimes multiple channels grouped by country or region) offering live streaming of their content. For the most part, though, this requires a device with **Adobe Flash Player** installed. If you don't want to install Flash on your device, your best bet is to use the **Skyfire** browser (see p.160) to access a channel's stream directly from its website. To watch UK TV, you can also try **MyPlayer**, which works without Flash, or for a selection of international channels, try **Online TV Player**.

Remote control

If you prefer the (relatively) big screen, and watch movies and TV at home with the help of a set-top box or multimedia PC, you can almost certainly find an app that'll turn your Android device into a Wi-Fi or Bluetooth remote control. Mac and iTunes users should be able to use the **iSyncr** and **Remote for iTunes** apps, while PC users can try **Universal Remote** as a general starting point. For something more specific to your set-up, just search the Market for the name of your set-top box or media player to see what's available.

MX Video Player/Pro Free/$5.62 (£3.44)   4.7

There are some exceptional video players on the Android Market, and MX Video Player is one of the best available. It'll happily play almost any video and subtitle format beautifully. Under the hood it supports multi-core decoding (taking full advantage of your device's dual core CPU) with highly optimized codecs and renderers. You can swipe left to right for forward and rewind and pinch to zoom in and out, while the edges of the screen control brightness and volume. For a great alternative, try **MoboPlayer**.

Mizuu Free/$1.65 (£1.16) 4.7

Tablets-only app that pulls together your movies into a beautifully displayed database, including summaries, actor information, trailers and posters from IMDb. Mizuu isn't a media player itself, but if you have a lot of movies it's a neat way to organize and browse your collection.

Plex $4.99 (£3.10) 4.0

Plex is a media player that works in conjunction with Plex media server for Mac or PC (available for free from plexapp.com) to easily browse and stream movies, photos and music directly from your computer to your Android device.

Movies by Flixster Free 4.6

Lets you read movie reviews and information from Rotten Tomatoes, get local cinemas and show times, along with nearby restaurants, watch trailers and share your own ratings with Facebook friends, manage your Netflix queue and more. Want more? Try the **IMDb** app.

Music

Getting music onto your phone or tablet can be done in a variety of ways: you can transfer files over from a computer via a file manager and USB cable (see p.105), sync wirelessly using DoubleTwist and Airsync (see p.195), buy MP3s using a service such as Google Music™ (US only) or Amazon MP3, or stream from the web with Spotify, Pandora or their many equivalents.

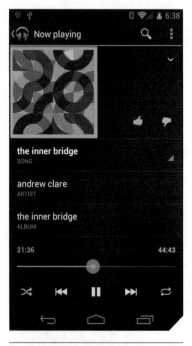

Well (cough), this looks like an interesting album! **Music**'s simple interface lets you get to your tracks with the minimum of fuss.

Google Music

Google™'s own **Music** app will probably have shipped with your device. It's a decent, straightforward MP3 player with most of the basic functions you're likely to need. You can browse your collection by artist, album, and so on from the tabs at the top of the screen, or just swipe between them.

Tap an item to play it, long-press to bring up other options, including adding to playlists, searching, and shopping for the artist online. (Currently, US users will be taken to the Music section of the Android Market™, while users based elsewhere get a Google search result via the Browser.)

As well as streaming to your Android device, you can browse, listen to and manage your Google Music collection from your computer's web browser.

Playback controls and the currently playing track details are also available from the notification bar and the lock screen, so you don't have to fumble around to skip a track, pause or see what just came up in your shuffle.

Store your collection in the cloud (US only)

When you buy songs or albums from the music section of the Market, they're added to your Google Music account and stored in the cloud. You can stream this music over an Internet connection to any of your devices with the **Music** app. You can also upload your entire music collection from your computer to Google's servers (currently you get storage for 20,000 songs) and stream these too. To do this, you'll need to download and install **Google Music Manager** to your computer and tell it where your songs are stored, then just leave it to gradually upload your music. Songs

Using Google Music services outside the US

Google Music™ services may well have rolled out to the UK and other countries by the time you read this, but if it's not available in your country, you can still get access to its cloud storage and streaming facilities with a little bit of skulduggery.

Basically, the process involves logging in to music.google.com with your Google account details using a web proxy to make it look as if you're logging in from the US and claiming one of the free songs on offer. Once you've done this successfully you can log in normally and still get access to the service. What you won't be able to do is buy music from the Android Market, but the storage and streaming facilities will be yours for the taking. A full guide can be found here: goo.gl/PLcN5

will be playable on your Android devices as soon as their uploads complete.

Tip: Similar cloud storage/streaming is offered by **mSpot**, or you can stream music directly from your computer with **AudioGalaxy** or **Subsonic**.

Making tracks available offline

It's amazing to be able to stream your music from the cloud, but there are times when it's more convenient to have albums stored locally so that you can play them without the need for an Internet connection. From the Music app, long-press an album and select **Available offline**. The album will download to your device and be accessible to play as a normal MP3.

To see which albums are available for offline listening (and avoid wasting time browsing through tracks you don't currently have access to if you're out and about), select **Offline music only** from

Similar services for buying music

The next most Android-friendly source for music after Google is **Amazon MP3**, which provides a similar service complete with cloud storage (again, for US users only), streaming and an integrated music player and MP3 store app. If you're a fan of **DoubleTwist** (p.194) check out the desktop version, which has Amazon's MP3 store built in, so you can buy tracks and have them sync directly to your device iTunes-style.

You could also try **7digital Music Store** from the Market, for another combined MP3 player and shop.

the menu, and all of your cloud-based albums will be hidden from view. It's also worth delving into Music's settings, where you can set the app to automatically hide unavailable music depending on Internet availability. While you're there, you can select **Download via Wi-Fi only** and/or **Stream via Wi-Fi only**, to avoid running up your remote data usage.

Just because you can stream music to your device, it doesn't mean you have to. Tracks purchased on Google Music are 320 kbps MP3s, so you can download them to your device and use it as a regular old-fashioned MP3 player.

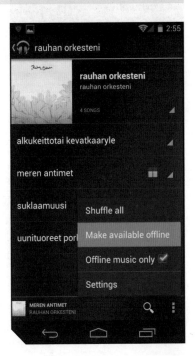

As a music player, DoubleTwist is a step up from the stock app, with rating support, and some nice navigation features like swiping to skip forward and back. Add in video playback, streaming radio, desktop sync and optional (rather expensive) add-ons, including automated album-art downloading, and it's tough to beat.

DoubleTwist Free   4.4

A media player for both your Android device and desktop computer. Once installed on both, you can use it for syncing of music, videos, photos and podcasts from one device to the other over USB. DoubleTwist on the desktop will update its media library from your iTunes or WMP playlists (or any other folders you specify). It offers a rudimentary media player of its own, although you may prefer to retain your existing set-up for actually playing music at the computer end. The desktop player has Amazon's MP3 store built-in, giving the whole set-up a very iTunes-like feel for buying and syncing music. It also has a facility for subscribing to podcasts, which also sync with the Android app. Meanwhile, the Android app provides a solid, all-round media player, with music, video and podcasts all covered. It has a streaming radio player, too, with hundreds of stations browsable by category. For a similarly featured syncing desktop and app combination try the excellent **Winamp**.

AirSync $4.99 (£4.99) 4.2

The AirSync add-on for DoubleTwist brings wireless syncing to the table. With DoubleTwist on your computer and AirSync on your Android device, it's as simple as dragging files from one to the other. Enter the passcode from your device into the computer, then, once the two are paired, your Android device shows up in DoubleTwist's sidebar, the same as if it were connected via USB.

If you don't want to mess around syncing individual albums, you can configure DoubleTwist to automatically sync everything you've got, or just your music, video, or podcast subscriptions.

AirSync in progress

Please wait while doubleTwist wirelessly syncs files with your computer. You can set up automatic sync for this device in doubleTwist on your computer.

Cancel Sync

Tip: You can use Windows Media Player (from version 11 onwards) to sync files with your Android™ device. Simply connect over USB, and on your PC click the **Sync** tab. You can drag music or videos into the tab area to queue them up and then press the **Start sync** button to copy them across.

PowerAmp $4.99 (£3.07)   4.7

Very powerful, dedicated music player. For playing locally stored files (streaming isn't supported) it's one of the best out there. It has an unparalleled 10-band equalizer and tone controls (savable to a pre-set library), built-in album art finder and tag editor, cross-fade and gapless playback support, lyrics support (with lyrics search), lock screen controls, an excellent set of widgets and themes, and has a range of add-ons available, including one which allows you to shake your phone to change tracks. It handles most music files, including MP3, MP4/M4A, alac, ogg, wma, flac, wav and ape, and supports scrobbling your listens to LastFM. Similar features can also be found in **Player Pro**.

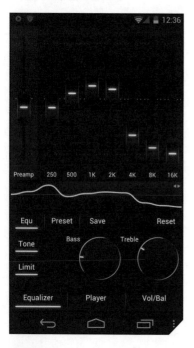

Subscription services

In case you hadn't noticed, the Internet is spilling over with subscription-based music streaming services. Pay a monthly fee and get instant access to millions upon millions of tracks anywhere, any time. Some, such as Pandora for US users, let you pick one album you like and proceed to conjure up a streaming channel of other stuff you'd possibly like. Most of these services (Pandora, Spotify, and Rdio, for example) now have apps that let you stream to your Android device; subscription fees vary, but many offer free trials.

TuneIn Radio/Pro Free/$1 (£0.69)   4.5

Offers more than 50,000 local and international radio stations and 1.2 million Internet streams. Ideal for finding new stations or taking your favourite ones with you when travelling. The Pro version also lets you record streams to your device's SD card.

Shazam/Encore Free/$4.99 (£2.99)   4.5

An ingenious app that will identify a song you play it and tell you everything about it, including the lyrics. The app links up nicely to other services so you can buy songs on Amazon, watch the video, listen on Spotify, buy concert tickets and more. If Shazam isn't doing it for you, try **SoundHound**.

Books, comics & magazines

Words and pictures

The number of different formats used for music and movies can be a little confusing, but the dizzying range of eBook formats available makes this look like a walk in the park.

Digital publishing seems to be carved up by the major retailers, each offering books in their own proprietary formats (Amazon's Kindle, for example) linked to the hardware eReaders they peddle. Add to this PDF, ePub, fb2, chm and any number of other book formats and things start to get a little overwhelming. And that's just books. Throw comics and magazines into the mix, and you're knee-deep in three-letter file extensions.

With a few well-chosen apps, however, you'll find that your Android™ device can handle pretty much anything you throw at it.

PDF readers

One of the most useful formats that will crop up time and time again is PDF. Your device may have shipped with a PDF reader already installed, and Amazon's **Kindle** app now offers rudimentary PDF support, but there are apps out there that will serve you much better for viewing PDF files. Features to look for include text reflow (allowing you to read the page at any magnification

ezPDF Reader includes more or less every feature you could need from a PDF viewer, including a night mode, which inverts text onto a black background and makes images grayscale. Easier on the eye, and also on your device's battery.

without scrolling around), bookmarking, annotations and search. If you don't mind forking out a few dollars you can pick up **ezPDF Reader** (p.152) or **BeamReader**. You'll also find solid PDF readers included in multifunctional office suites like **Documents To Go** (p.145).

Books

If you already own a Kindle, Nook, Kobo, or one of the other hardware book readers, your allegiances are probably already set. Your best bet, if don't want to juggle multiple accounts with other retailers, is to settle with the service you're already using, grab their app from the Android Market™ so you can access your existing library with your Android device, and keep a good all-rounder like **Aldiko** (see p.201), **FBReader** or **Mantano Reader** to hand for everything else.

Google Books

As with the **Videos** and **Music** apps, **Google Books**™ is a simple app that integrates with the Android Market to browse and make purchases. Buying a book stores it in your cloud-based Google account, and the Books app opens it from there. Your books can be browsed in a list view or a pretty scrolling carousel. Open a book by tapping on it, and swipe to change pages. Tapping on a page brings up an Action bar at the top and a progress bar at the bottom that you can drag to go straight to a specific page. The app will save your last position and open the book at that page the next time you access it. You can also jump to chapters by selecting **Contents** from the menu.

Like Videos and Music, you can pin individual items for offline access by selecting **Make available offline** from the menu and tapping the buttons of the books you'd like to store locally. One

Google Books, possibly a little light on functionality but heavy on the eye candy, including this 3D carousel view, here shown displaying four classics from our collection.

of the app's more interesting features is the ability to view books as scans of their original pages: select **Original pages** from the menu, or switch back to **Flowing text** once the novelty has worn off. This seems like a bit of a gimmick but it's actually pretty useful when viewing books with illustrations or diagrams.

Aldiko Free/$2.99 (£1.99) 4.4

Beautifully crafted book reader that lets you download from thousands of free public domain and Creative Commons titles, as well as buying current releases. Standard features like bookmarks and swipe navigation sit alongside useful additions such as swiping the screen edge to adjust brightness. You can also import your own ePub books or add URLs to search other online catalogues. Useful add-ons from the Market provide sync support. Downloaded books appear on the shelves of a nice wood-effect bookshelf for your perusal. Pipe and slippers optional.

Mantano Reader Free/$7.49 4.6

Lightning fast with a ton of features, Mantano is a PDF and ePub reader that looks great on tablets. Facilities include customizable themes you can apply to books; a well laid-out, searchable library, allowing you to add tags and organize books into categories; text highlighting; shareable quotes and annotations; text-to-speech; dictionary search; an excellent range of view modes for PDF, and more. For other formats, your best bet is the free **FBReader**.

Comics

Comics come in a variety of formats, from the cbr, cbz and cb7 eComic formats to good old PDF and zipped (compressed) archives of standard image formats such as gif, jpeg, and png. A good comic viewer will cope with all of these, and for the most part you won't have to worry about which formats your comics arrive in.

For a capable, free, lightweight viewer, try **Droid Comic Viewer**, **Comica Lite** or **Perfect Viewer**. Tablet users will appreciate the larger screen-optimized features offered by **ComicRack** and **Komik Reader**.

Comics Free 4.1

Comixology's combined viewer, library and store app gives you access to its huge digital library from publishers including Marvel, DC, Image, IDW, Archie and more. The viewer includes useful small screen optimizations.

Magazines

Magazines are beginning to creep their way into the digital realm, now broadly supported across other platforms and with support beginning to emerge for Android™. Aside from the ubiquitous PDF format, you can now use a virtual newsstand service like Zinio (below) to buy and read digital magazines.

Zinio Free   3.7

Beautiful, easily navigable reader and store for digital magazines (some with exclusive multimedia content) from major publishers around the world. Buy single issues or take out a subscription and sync your collection with other devices. Read from the original full-colour layouts or in text-only mode. The app looks and feels amazing on both phones and tablets, and comes loaded with a selection of free editions covering a range of interests to get you started.

News & podcasts

Grabbing your daily fix

Many websites and blogs add any new content that they publish to a "feed" (usually RSS or ATOM, but you don't need to worry about the distinction) that you can subscribe to anywhere you see this symbol (above, left) on the Internet. A news reader or aggregator will collect new articles so that you can read them all from one app without having to trudge round the same old sites every day. Podcatchers operate in a similar way for podcasts, routinely checking for new episodes and downloading them for listening (or watching) at your convenience.

As usual, Google™ offer their own simple but effective free apps for aggregating this web content: **Google Reader**™ and **Google Listen**™. Reader lets you subscribe to news feeds, syncing with a Google Reader account for easy subscribing from a web browser; while Listen can subscribe to and play podcasts (these subscriptions are also manageable from Reader). If you're serious about your web content though, try out **Pulse** (opposite) or **News360**.

At the time of writing, Google had just introduced **Google Currents**, a new (currently US-only) service serving up free content from publishers such as Forbes, Techcrunch, Popular Science, Saveur and many more. Currents also integrates with your Google Reader subscriptions and presents them in a beautiful magazine-like format alongside "trending editions", an hourly edition covering the latest stories in a range of categories.

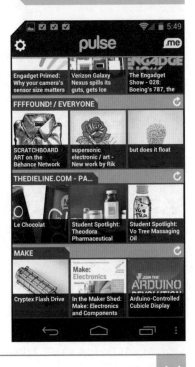

Scrolling around Pulse's responsive mosaic of pictures and text is incredibly satisfying. It's also a great way to grab all your daily news for reading on the train to work. Another exquisite news reader worth a look is **News360**, available in separate flavours for phones or tablets.

Pulse Free   4.4

Pulse lets you set up RSS feeds to your favourite sources and stores them for you in a friendly, scrollable grid format. You can set up different pages for different kinds of content and adding or removing feeds is a doddle. It'll even pull in content from your social networks, allowing you to wallow through your morning coffee / Facebook / news ritual all from the same location. It syncs with a Google Reader account, if you use one. If you sign up for a free pulse.me account you can also sync with Read It Later, Instapaper and other services. Some newsworthy alternatives include **Newsrob**, **gReader**, and **Feedly**.

DoggCatcher's edge over the competition lies in its configurability, including a range of useful options such as the facility to download new episodes only over Wi-Fi , or only at night, or when your device is otherwise inactive. Its developers are also very responsive to feature requests from the user community, issuing updates above and beyond the call of duty.

DoggCatcher $4.99 (£3.22)/Free 7-day trial 4.5

Everything you need in a podcatcher, DoggCatcher manages podcasts (both audio and video) and news feeds. It supports automated downloads and deletion of played files, playlists, variable playback speed (with the **Presto** add-on), and playback position recall. It comes with a huge library of feeds to peruse, and can provide recommendations based on your subscriptions. For a prettier interface in exchange for a few less high-end features, try **Pocket Casts** for $2.99.

Tip: DoubleTwist's desktop app (p.194) can subscribe to podcasts and sync them to play with DoubleTwist's integrated podcaster on your phone or tablet.

Games

Fun fun fun…

Developers have moved quickly to show what's possible with the Android™ platform, and the new hardware acceleration in Ice Cream Sandwich allows for massive increases in gaming speed. The ever-expanding gaming section in the Android Market™ is helpfully browsable by category (Arcade & Action, Brain & Puzzle, Cards & Casino and more).

The Market hosts tens of thousands of games – many of them free – in every conceivable genre, from puzzles and brain teasers, to strategy to arcade and action games. Buying a game from the Market is pretty much the exact same procedure as for buying an app (see p.71). To help you sift through all this stuff and pick yourself a winner, check reviews of all the newest and coolest games at the following sites:

- ▶ **Droidgamers** droidgamers.com

- ▶ **Eurogamer** eurogamer.net

- ▶ **Best Android games** bestandroidgameaward.com

- ▶ **Android games hub** androidgameshub.com

- ▶ **Gamespot** gamespot.com/android

- ▶ **Pocket gamer** pocketgamer.co.uk

You can play mobile-optimized Flash games at android-games. net **and** kongregate.com **(see over).**

Get six hundred games for free.
Kongregate has enough variety to keep
you coming back again and again.

Kongregate Arcade Free 4.6

Flash-based gaming site Kongregate now has an Android app
that gives you access to their mountain of free games – more
than six hundred at the time of writing – more than enough
to keep you busy. Create an account and you can compete
with friends on leader boards, rate and review games, collect
badges and track your scores. There's a strong social element to
Kongregate, and the mobile app gives you full access to your
profile, wall and inbox. Or you can just fire it up, pick a game
and start playing. Games load up quickly and run full-screen.

OnLive looks set to be an awesome gaming platform, with hundreds of games already available from its rapidly expanding library. You can rent titles for a few days, buy outright, or take them for a quick test run.

OnLive Free 4.2

A cloud gaming service where you can buy or rent console and PC games then stream them to your phone or tablet (or your Mac/PC/games console). Games vary in price but by downloading the OnLive app you get a free copy of Lego Batman (above) to get you started. Games are cross-platform, allowing you to pick up a game in the evening on your TV that you started on the train to work using your phone. You can play using on-screen touch controls, or buy OnLive's wireless controller from onlive.com/store.

Is Joel Smith happy working in the video rental store? Does Macy Spacey like the new paint job down in the lobby? Before you know it you'll be sucked in to the world of your tiny tenants' tiny minds.

Tiny Tower Free

4.5

Adorably cute and strangely compelling little game where you build and manage a tower block. Build apartment and retail floors, wait for the tenants to arrive and assign them jobs based on their interests. Make sure your businesses are well stocked to keep the money rolling in so you can build more floors. If your tenants don't like the job you've given them, or are being just plain difficult you can reassign or evict them and hope someone more appropriate moves in. Meanwhile, keep an eye on the elevator to ferry customers and visitors up and down between floors.

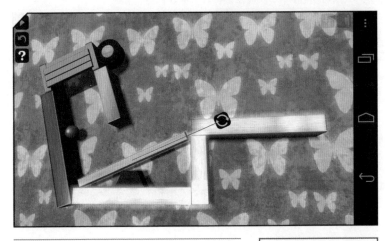

Standard mode (above) presents a series of conundrums to be solved with a fixed set of pieces. Completed levels can be replayed in free-build mode, which lets you modify the puzzle. Sandbox mode takes this a step further and allows you complete freedom to combine parts into any kind of puzzle or thingamajig your imagination can muster. You can then upload these to the community area, alongside some breathtaking inventions.

Apparatus/Lite $2.45 (£1.80)/Free 4.6

Apparatus is a beautifully rendered construction puzzle where the objective is to coax the little blue ball into the little blue box. You achieve this with an array of cables, cogs, generators, ropes, buttons, ball-bearings and other items. As with most Android games, it starts you off with a couple of no-brainers and then hits you with a difficulty curve that takes you from beard-stroking through to head-scratching and ultimately sleep-depriving complexity, each new level requiring you to work with bigger and weirder contraptions to achieve the solution.

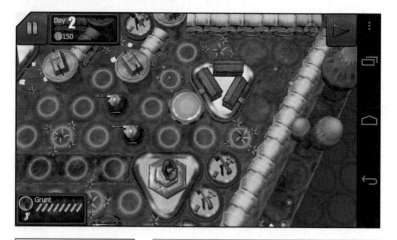

As you play through the game's thirty missions – ranging from simple kill-the-enemy objectives to more involved defence and capture scenarios – you unlock specialized units and promote the ranks of your existing battalion. There are also a few different game modes, including a two-player option that lets you pass and play on the same device.

Great Little War Game $2.99 (£1.86)   4.6

Mobilize your troops through a series of increasingly complex battlefields in this turn-based strategy game, dispatching enemies as they approach. The game plays like action chess, and you'll have to think ahead if you want to use the varied terrain to your advantage, setting up ambushes and tactically exploiting your enemy's weaknesses. Take turns with your opponent by tapping the worryingly rotund troop, tank or boat you'd like to move or attack with, pick a destination or target and watch while another piece of the incremental action plays out.

Can you dig it? The pocket edition is currently missing most of the features of the desktop game, but it's still compelling to just worm your way around under the ground for hours on end. A free demo version is also available.

Minecraft – Pocket edition $6.99 (£4.29)

4.3

Minecraft drops you into a strangely beautiful, blocky three-dimensional world. There's no real objective or direction in the game; you just sort of wander around, collect stuff, make stuff and dig. Punch a few trees down and collect enough wood to make an axe; from there you can start to dig down deep into the ground, collect other minerals for building other stuff, and construct anything from huge cities to complex mechanical traps.

Behind its seemingly basic gameplay, Flick Golf! is devilishly addictive, so if you're downloading this game, make sure you don't have anything important you're supposed to be doing any time soon (like writing a book, for example).

Flick Golf! $0.99 (£0.69)

4.2

Here's your golfer, here's a golf club, here's a golf ball, there's the hole, now get to work! It's a simple premise for a game made even simpler by Flick Golf!'s intuitive controls: swipe the screen to swing and hit the ball. Once your ball is in flight, you can affect its trajectory by swiping a bit more demonstratively in the direction you'd like it to travel. The immersive 3D environments help draw you in and the underlying physics engine makes gameplay very responsive and natural while presenting just the right amount of difficulty to keep you coming back for more.

travel & navigation

Travel

The A–Z of A to B

They don't call them mobile devices for nothing. Wherever your travels take you – whether it's halfway around the world to swim with dolphins or all the way around the corner to buy some milk – you'll find the location-aware facilities provided by your Android™ device invaluable.

Android phones (and many tablets) have a built-in GPS (Global Positioning System) chip, which **Google Maps**™ and other apps can use to provide you with turn-based navigation, geocaching (tracking your position over time), information about local services and more.

Google Maps

Wherever you go, if your device is equipped with GPS you can instantly find your location on a Google Map. Maps is one of Google's most honed products, and the mobile version is constantly being updated with new features. Recent additions include 3D renditions of some major cities and indoor maps of large structures like shopping malls and airports.

Google Maps is the hub for a family of location-oriented apps that also includes **Latitude**, **Navigation**, **Street View** and **Places**, all of which you can access from the drop-down menu at the top of the screen.

Basic Maps view in portrait mode. The Traffic layer is enabled, showing traffic flow in colour.

Maps' initial view shows your general location on a standard street-map, with you at the centre, represented by a blue arrow which roughly indicates the direction you're facing. Maps acquires this information from a range of sources: primarily GPS if you're outdoors, but also from mobile phone masts, and your device's barometer if it has one.

Moving around the map is a simple case of swiping around with your finger. Pinch the screen to zoom out and double-tap or unpinch to zoom in. You can also rotate a map using two fingers (see p.49). To get the map back to your current location, tap the ◉ crosshairs (top right), or to return a rotated map to orient North, tap the red needle (shown in top left of picture opposite).

Tap the crosshairs twice for an aerial view; the landscape appears spread out before you and rotates according to the direction you're

Same location, different tag. Standard long-press (left), and a single tap with **Bubble buttons** enabled (right; see tip, opposite).

Aerial view and satellite layer enabled. Notice the red compass needle has appeared (top left).

facing. You can exit this mode by tapping it again, or by tapping the red compass needle.

Long-pressing on the map brings up a tag showing the address at that location. Tap the tag for more options, including searching for nearby amenities, calling businesses, Street View and more. You can tap the star in the top corner of this page to save the location to My Places, a list of favourites you can return to later.

> **Tip:** Bubble buttons is a useful add-on you can find hiding in **Menu > Settings > Labs**. When it's activated, you can tap an address for quick access buttons to navigate there, or, if you've tapped on a shop or business, to phone them directly.

To see more information on your map, tap the 🔶 layers button in the Action bar, to toggle other layers of information, including live traffic information, which colours parts of the road green, yellow or red depending on congestion levels (with alarming accuracy).

Other buttons in the Action bar provide shortcuts for Search, Navigation (p.221) and Places (detailed information about local amenities, see p.226).

Saving maps for offline viewing

To avoid using mobile data to see a map when out and about, you can "cache" map areas for offline viewing. Select **Menu > Settings > Labs**, where you'll find some optional add-ons. Tap **Precache map area** to select it, and then exit back out to the main Maps view. Long-press somewhere roughly in the centre of the area you'd like to cache and tap the address tag that pops up. You'll be taken to a page of options for that particular address, select **Precache map area** and Maps will download map tiles for

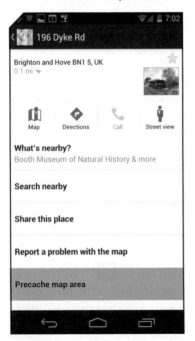

everything within ten miles (it just downloads the basic street-map tiles, not the satellite images, traffic information or information on other layers). You can Precache as many areas as you like, and manage them from **Menu > Settings > Cache settings > Precached map areas**. Precached maps are handy for finding your way around in a new town, but, annoyingly, you'll still need a data connection to search for an address or use navigation.

Tapping on a tag (see p.219) brings you to a page like this.

Choosing an alternative route from the three available options for this particular trek.

Navigation

Maps provides SatNav-like turn-based navigation from a separate app called, you guessed it, Navigation.

You can access Navigation directly from your app tray or from its button in car mode. Simply enter your destination (either spoken, typed, or selected from your contacts or starred places) and mode of travel (by car or on foot; you can change this later from **Menu > Set destination**) from the drop-down list (top left).

Tip: You can also perform simple voice searches, such as "Navigate to pizza" and select your destination from local search results.

The app will churn for a bit while it finds your GPS location and directions. While it's doing this you can tap 🔀 or **Menu > Route info** to see details of the route, and from there tap the 🔀 button to see alternative routes for your journey (selectable from the top of the

screen), or tap ⚙ to specify route options (avoiding toll roads or main roads). Once you've settled on your route, tap ☰ to see the directions as a list, or tap ⚊ to start navigating. The app will enter a SatNav-like mode and start speaking directions at you and updating the screen as your journey progresses. Toggle between an overhead view and an in situ view with the compass ⬈.

You can also tap 👤 for a photographic street view. Tap it again to enable looking around by swiping around the screen.

Selecting **Menu > Layers** lets you toggle various layers of information, similar to those found in Maps, but more journey-specific (restaurants, petrol stations and so on). There's also a Satellite option here, but unless you have unlimited mobile data, we don't recommend switching this on for an actual journey.

Tip: Add a shortcut to your home screen for finding your way home or back to your hotel. From the app tray, select the **Widgets** tab and find **Directions & Navigation**. Long-press to add it to your home screen and enter the destination and preferred mode of transport. Now, with a single tap you can get directions from anywhere back to this address (but so can anyone else who picks up your phone, so it may be wise to set an address a block or so away from where you actually live).

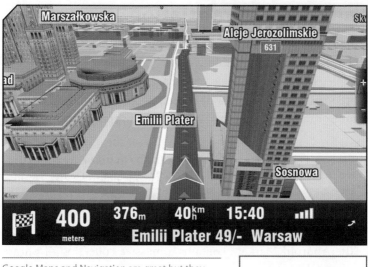

Google Maps and Navigation are great but they rely on an Internet or data connection in order to fully function. For a full-featured SatNav, using the latest TomTom maps, Sygic is an affordable solution.

Sygic from $10.00   4.0

Sygic is a full offline turn-by-turn voice-guided navigation app that transforms your device into a fully equipped SatNav. It has more features than we can list here and looks incredible, with amazing 3D rendering of cities and landscape. Maps are available to buy in a range of country or continent packs. Before you jump in and sign up with Sygic, check out the competition from **CoPilot Live** and see which feature set suits your needs.

Language & translation

When travelling abroad it's always useful (but not always practical) to have a basic understanding of the local lingo. Whether you're completely unable to speak a language or your vocabulary is just a little rusty, a couple of well-chosen apps stashed away on your phone or tablet will help make your time away a lot easier.

Tip: Snap lo-res photos of essential items before you go on your travels and store these in categorized folders for food, household items, amenities, etc in your Gallery (p.178). Now you can just show someone the picture on your phone and the only phrase you need is "Where?"

The Android Market™ contains a massive collection of phrase books and language tutors of varying quality, as well as some excellent live translation apps that let you speak a phrase and have the translation spoken back to you. The downside of these text-to-speech and speech-to-speech translators is that they require an Internet connection, something you may not always have affordable access to when overseas. Your best bet in these situations is to have a language-specific offline translator pack, such as those offered by **BitKnights**, or to use a text-based offline translator such as **Tourist language learn & speak**.

QuickDic Free   4.5

Indispensable translation dictionary, QuickDic is fast, ad-free and stores dictionaries on your device's SD card for offline translation. Converts English to 50 or so other languages.

Tip: Google Goggles™ is an awesome image recognition tool but you can also use it to get instant translation of foreign-language text – just point and shoot.

Google Translate Free 4.6

Google Translate™ provides fast and reasonably accurate speech-to-speech translation, but in some languages it can speak back at you a little too fast. If you find yourself constantly wanting to slow it down a bit to understand individual words, try switching to **Trippo Mondo**, which lets you control the speed of spoken translations.

There is a penguin in my bath

ペンギンはお風呂にあります

Where can I buy a gun

どこで銃を購入することができます

日本語で続きを話す

ペンギンわどこですか

Where is the penguin

Reply in English

Type or speak your phrase in any language and have it translated into any other. For snappier, real-time translation between two tongues, switch to Conversation mode. Translate also integrates with your sharing options, allowing you to send text to it from other apps without copy or pasting, and an SMS button selects directly from your text messages. A send button will also copy your translation back out to a variety of sources. The closest thing to a universal translator currently available.

Travel guides and news

Whether you're scoping out places of interest on the other side of the planet or just looking for a good place to eat in the next town, the Internet is riddled with sites ready to help you make an informed choice. If you're headed for a major city, look it up in the Android Market™ and you'll find a slew of city guides, subway maps and entertainment listing apps all downloadable to your Android™ device.

Rough Guides offer excellent city guide apps for London, Paris, New York and Rome, or check the lists available from **TripAdvisor**, **DK Travel**, **GuidePal** and **Triposo**, all offering interactive maps, attractions by category, local histories and more.

Google Places

Google Places™ is part of the Maps family of apps; fire it up from a selected Maps location for all kinds of local information: restaurants, cafés, pubs, entertainment, ATMs, hotels, taxis and more. Select a category and peruse the reviews (or add your own), get directions, see photos, or call them up. It also links to other social reviewing networks (Qype, beerintheevening, and plenty of others) for a broader overview of the establishment in question.

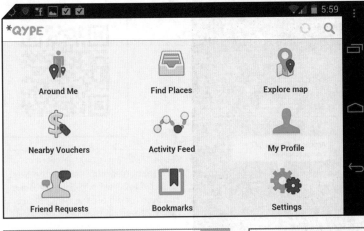

Qype Free 4.1

Qype provides a kind of social network element to social reviewing. You can set up a profile and make friends if you want to, or just ignore that side of it and use the invaluable real-life reviews from real-life customers. The Find Places section gives you much more navigable paring-down of your options than Google Places, based on what you're actually looking for. It also offers discount vouchers for participating local businesses.

Alfred Free 4.4

Make friends with this clever little chap to get curated recommendations. Alfred uses an artificial intelligence engine to figure out where you should go next.

FlightTrack covers 1400 airlines and 4000 airports worldwide. Whether you're travelling yourself or picking someone up from the airport, this handy little app will make sure you're always up to date on a flight's progress (or lack of it).

FlightTrack $4.99 (£2.99)

4.4

Essential for anyone getting on a plane, FlightTrack gives you real-time flight tracking from your mobile device, with departure and arrival information (updates are sent to your notification bar), zoomable live tracking maps with weather imagery, info on delays (with detailed forecasts), gate numbers, cancellations and help finding an alternative flight. You can share your flight status via SMS, email, Twitter or any other messaging app. It'll even check your seat number against SeatGuru's plane seating maps. For a free alternative, try **FlightView**. Need a hotel when you arrive? Try the free **Booking.com** reservations app.

Öffi – Public Transport Free 4.6

Impressive public transport timetables, with live arrival and departure details covering most of the UK and Europe, as well as some locations in Australia and the US. When viewing a list of local departure points, each has a live compass showing its direction, great for finding a bus stop or station in a strange city. If your destination isn't covered, note that public transport maps and apps for most major cities are available for free from the Market.

WeatherBug Free 4.1

No need to stick your hand out the window ever again: WeatherBug has everything covered – UV data, pollen count, temperature, lightning strikes, humidity, air pressure, wind speed, cloud cover – anything you could possibly want to know. For hour-by-hour forecasts in less detail, try **The Weather Channel**.

Trip Journal's leathery, yellowed interface adds a scrapbook-like finishing touch to a well thought out and easy to use, media-rich travelogue.

Trip Journal Free/$4.99 (£2.99)   4.2

A great way to share your travel experiences in real time with friends and family. Trip Journal tracks and maps your location while you upload photos, video and journal entries, and makes it all instantly available through your social network of choice. It integrates with major online services like YouTube™, Picasa™, Flickr, Twitter, Facebook and Google Earth™ and lets your friends view your progress and make envious comments. The Lite version gives you one trip for free, or you can buy additional trips in packs of three for 99¢ cents, or activate unlimited trips permanently for $4.99.

security
& privacy

Security & privacy

Playing it safe

The number of Android™ devices in use is skyrocketing, and wherever there are masses of people using a service, there will always be a contingent looking for new ways to exploit those masses. While actual viruses may not be prevalent on the Android platform, there is an increasing risk from malware and malicious apps that you should be aware of.

The Android Market™ is not heavily moderated, but it's been documented that one or two apps (out of thousands) have been removed for attempting to harvest users' financial data. At present the risks are minimal, with any vulnerabilities quickly patched up by Google™; still, it seems only a matter of time before malware becomes just as prevalent on this platform as any other.

As well as the software risks associated with using any kind of computing equipment, smartphones and tablets are portable devices that often contain sensitive personal data, carrying with them a whole other level of security risk due to loss or theft.

Apps and security

"Permissions" and how they work

When an app is installed it provides a list of its "capabilities" to the operating system, basically all the different functions it will need to access. These will be shown as "permissions" for you to

Security isn't just about your phone or tablet

It's worth remembering that while your phone or tablet itself may be unscathed by a computer virus, it could transmit viruses between other devices (someone else's computer and your own, for example) in much the same way as any other storage media can. So if you routinely connect your device to more than one computer, take the same precautions you'd take with any other portable storage (external disks, memory sticks and so on) and use the computer to virus scan any files you're moving to and from the device.

read through before you go ahead and install the app.

Once installed, it's impossible for the app to do anything (such as using your phone to make calls or access your GPS location) that it hasn't declared in its capabilities. There's always a risk, however, that some apps could declare capabilities beyond those they'd legitimately require, potentially opening up some worrying security or privacy issues. It's worth scrutinizing the permissions that you're granting any application to make sure it's not asking to do anything you'd consider unnecessary.

Tip: Permissions for your already installed apps can be seen by going to **Settings > Apps**, tapping on an app and scrolling to the bottom of its info page. Take a look at some of the applications you know and trust to see what kinds of permissions they require in order to operate.

How to tell whether permissions are legitimate

Use a degree of common sense: does that shoot-'em-up game you're about to install really need access to your contacts list? If in doubt, do a quick web search and see if anyone seems to be ringing alarm bells about that particular app.

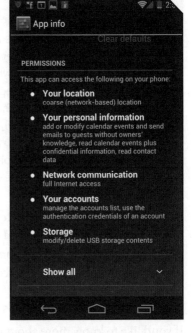

A typical permissions list, this one is for the **Astrid** app. It can access your account authentication details, but it would legitimately need those in order to sync with your Google Tasks list.

Just to confuse things, sometimes an application may request permissions that seem unreasonable but are necessary for it to function properly. For example, a media player may need to monitor your phone calls so that it knows when to put music on hold for the duration of a call.

If you want to know more specifically which services your apps can access, use **Appshield** ($2.50) or **Permissions Denied** (free) for a level of detail not seen in Android's own application settings page, and the ability to selectively disable permissions you don't want to give any particular apps.

Anti-virus programs – do you need one?

It's practically impossible to build a virus for Android that would damage the actual operating system, so you may think an anti-virus app for your phone or tablet is unnecessary. However, they often incorporate other useful security tools, such as the facility to remotely lock the phone or check installed programs against lists of known malware. Because the only way malicious software can currently get onto your phone is by you actually agreeing to install

Malicious QR codes

A recent threat to emerge is the use of malicious QR codes. When scanned, these connect your device to a website which can install trojans or other malicious apps, which, for example, use permissions to make calls or send text messages to covertly contact a premium rate number which is billed through your existing carrier. You wouldn't be aware of this until you get an astronomical phone bill at the end of the month. You're safe scanning the codes in this book, or from any other reputable location, but exercise caution if you're not certain that a QR code comes from a trustworthy source.

it, your best line of defence is to be very careful about what you install, be wary of unrecognized email attachments, always check the user reviews and permissions before you install an app, and only use apps from trusted sources.

Google's Good To Know site

If you'd like to know more about your data both offline and online, and how to protect yourself from the various risks associated with it, Google™ have put together Good To Know, an easy-to-understand

Good to Know

Stay safe online	Your data on the web	Your data on Google	Manage your data
... tips and advice for staying more secure on the web	... and how it makes websites more useful	... and how it makes Google services more useful	... and what you share with websites and Google

site which helps give you a deeper understanding of the main issues and offers up-to-date advice on how to stay safe.

▶ **UK** google.co.uk/goodtoknow

▶ **US** google.com/goodtoknow

Locking and unlocking your device

Perhaps the biggest risk – as with any device as portable as a phone or tablet – is that someone could steal it and access your private data. If you're worried about that possibility, the easiest line of defence is to set up a screen lock.

Open up **Settings > Security > Screen lock** in order to specify your preferred method for unlocking your device, from the basic **Slide** (p.45) to secure options like **Pin**, or **Password**, which prompt you for a number or word which you'll need to re-enter every time you turn your screen on. These can be a little cumbersome, so you may prefer to go with one of the snappier options: **Pattern Lock** or **Face Lock** (see below). Whichever option you use, once set you'll need to perform the unlock again when changing it to something else.

> **Tip: Settings > Security > Owner info** gives you the option to enter some text that will appear on your lock screen. If you lose your device, this could be used to help anyone who finds it get in touch with you.

Pattern lock

A pattern lock is probably the fastest and most straightforward way to secure your device. Select **Pattern** from the screen lock options and you'll see a square of nine dots. Draw a continuous line between the dots in a pattern that's complex enough to not be immediately obvious – not, then, the big "Z" that's used for demonstration

It's amazing how quickly your finger will remember a fairly complicated pattern like this, but if you decide to go for something really elaborate it might be an idea to draw yourself a little diagram for reference.

in the manual – but simple enough to remember. You'll then be prompted to confirm your pattern by drawing it one more time. Now, to use your device after it goes to sleep or is switched off, you'll be required to draw that same pattern to unlock it.

Face Lock

Face lock lets you unlock your device just by pointing it at your face. It's not as secure as the other options and someone who looks a bit like you or has a photo of you handy could potentially unlock your device. You'll need a phone or tablet with a front-facing camera in order to use it, and it can be a little temperamental, so you'll need to enter a pattern or pin lock as well,

for instances when face lock just can't recognize you (after a haircut or horrifying accident, for example). Select **Face Unlock** from the screen lock options and follow the instructions on screen.

What to do if you're locked out of your device

Enter the wrong code five times and you should see an option to set a new passcode. From here, you can enter the details of the Google™ account associated with your device (assuming you can remember them). If all goes well, you'll be prompted for a new passcode (write it down somewhere safe this time) and all should be right in the world again.

If all else fails, as a last resort you can point your computer's web browser to the Android Market™ website (see p.77) and install an app to your device called **Screen Lock Bypass**. As the name suggests, this will bypass the lock screen, allowing you to access your stuff again. The app won't actually fix the fact that you have an unlock code you don't know on your phone or tablet, and keeping it on your device long-term and never having a lock screen again isn't a good option, but you can use the opportunity to back everything up (see p.240), wipe the device clean (see p.36) and start again.

SIM lock

Phone users can set a separate SIM card lock to prevent other people using your handset. Select **Settings > Security > Set up SIM card lock** and then check **Lock SIM card**, and you'll be prompted for your existing PIN. If you don't know it, check the documentation that came with your phone to see if it's listed anywhere, or call your network provider and they'll provide you with it after verifying your identity. Once your SIM PIN is set up, anyone trying to use the phone will be required to enter this four-to eight-digit number. After three wrong attempts the SIM will be locked and you'll need to contact your provider again for a PIN Unlock Code (PUK), or buy a new SIM card and start again.

Backing up your data

Stuff like your Gmail™, Market™ apps, Google Calendar™ and contacts list all sync with the cloud anyway, so there's no need to back them up. Most of your other data (web bookmarks, saved games, app data, Wi-Fi login, internal settings, etc) can be backed up automatically to your Google™ account.

When you first use your device you should be presented with an option to **Back up my settings**. To check whether or not you have this enabled, head to **Settings > Backup & reset** and see if the **Back up my settings** and **Automatic restore** options are checked. If you wipe your device or set up a new device with the same Google account, your apps, data and settings will automatically download from Google's servers to the new device.

For backing up larger files, photos or video, your best bet is to manually copy them over to your computer, or use a cloud-based service like **Dropbox** (see p.106). Some apps will automatically back up your photos and video to Dropbox, including **DropSnap** and **Titanium Media Sync**. Root users (p.125) have some more comprehensive one-stop solutions for backing up everything in one go, the most popular being the popular **Titanium Backup Pro**.

MyBackup/Pro Free/$2.99 (£1.91) 4.3

If you want to back up more than just the essentials, your best bet is to use RerWare's MyBackup Pro. This allows for heaps of data, including SMS, MMS, call logs, your custom dictionary and other settings, apps, alarms etc to be backed up to your SD card or on RerWare's own servers.

Tip: The Android Market™ can restore any apps you've installed or purchased, but if you want to keep earlier versions of apps for insurance against newer versions not working, you can also use a file manager like **Astro** or **ES File Explorer** (p.108) to back them up.

Protecting your privacy

With any device that connects to the Internet, it's worth paying close attention to which elements of your online presence are being shared beyond your real-life social circle. Facebook, Twitter and other social networking services make it very tempting to update the world with anything from where you're staying on holiday to your innermost thoughts. You can amend privacy settings so that only your friends and family can see anything but the most basic details; but it's still safest to assume that anything you share online is accessible to far more people than you'd expect, perhaps including exes, future employers and cybercriminals.

All of this is even more important if using your phone or tablet to connect with GPS-enabled social networking services like **Foursquare** or other social networks that include a facility to "check in" to your location, as anyone resourceful enough can use a service like this as a springboard for accessing any information about you that's publicly available, as well as to track your location with the same degree of accuracy that your own phone can.

Tip: Bluetooth can be used to transmit data between your phone and other devices up to around 30 feet away and could be vulnerable to hackers, so unless you're actively using it – especially when out & about – switch it off (this will save battery power, too).

Safety when connecting to public Wi-Fi hotspots

There's a small risk with any Wi-Fi capable device that someone could be snooping in on data travelling via an open network. You should apply the same common sense when using public Wi-Fi from your Android™ device as you would were you accessing it from a laptop: unless you have to, avoid using Internet banking or any other service that requires you to enter sensitive personal data.

> **Tip**: You may want to uncheck the **Enable Web History** and **Stay signed in** boxes when setting up your Gmail™ account if you're concerned about privacy or if using a shared computer.

LastPass $1 (£0.69)/month | 4.5

If you're wise, you'll have different passwords for all the accounts you have dotted around the Internet. LastPass is a free password manager and generator allowing you to keep all your logins, passwords and PINs in one secure, encrypted database and access it with a single master password, it syncs well with other devices and has a plugin version for most major browsers (including Dolphin, p.161). For a free solution try **KeePassDroid**.

How to be sure your personal information is deleted when you sell your old phone or tablet

It's worth considering what kinds of personal data your device may still contain if you decide to donate, sell or throw it out at some point in favour of a newer model. You'll probably want to wipe certain information, but unfortunately it's a little more complicated than

Carrier IQ

Carrier IQ is hidden diagnostic software that some network providers pre-install on their mobile devices. The app gathers information on your location and call activity, ostensibly in order to improve network coverage. While Carrier IQ isn't malware and won't harm your device, it's difficult to ascertain whether it's installed and exactly what kinds of private information it may be harvesting. Currently, the general consensus is that it can report the following types of information:

▶ The number dialled when making phone calls, but no conversations, emails or SMS messages.

▶ GPS location.

▶ Any URLs visited from your device's web browser, but no actual content from those web pages.

To find out whether your device has Carrier IQ installed, you can download Lookout's **Carrier IQ Detector** app from the Android Market™. Unfortunately, even if you discover it's installed, you won't be able to uninstall it without rooting, which carries risks of its own (see p.125). There are a few apps on the Market which claim to be able to routinely kill the Carrier IQ process when detected – **Anti Carrier IQ** is the best option at the time of writing – but use it in conjunction with Lookout's aforementioned app to see if it's actually doing its job.

selecting files and telling the device to delete them. Deleting files in this way only reallocates the space they occupy as empty – ready to be overwritten if that space is needed – but the files themselves are still there and can be recovered using the right tools.

It is possible to permanently erase sensitive data so that it's rendered unrecoverable. Check to see if there are instructions in your device's manual or on the manufacturer's website. If not, try these data erasing instructions: tinyurl.com/2f4ue4.

To securely wipe an SD card (if your device has a separate one), head to **Settings > Storage** and select **Unmount SD Card**. Next, hit **Format SD Card**. This formats the card but doesn't actually wipe the data; to do that you'll need to plug your device into your computer to access the SD card and fill it up with large files (as near to full as you can get; music or movies are a good bet as they take up a lot of space) and then unmount and format again.

Once you've done this, head to **Settings > Backup & reset** and select **Factory data reset** to erase all your other data and settings, including whatever's stored in your device's internal RAM. Like the SD though, your RAM still retains a lot of this information even when deleted, so if you want to be really thorough, download and install as many large apps and games as it takes until your device won't accept any further downloads, and then perform the **Factory data reset** again. You'll need to install apps from outside the Android Market, however, to avoid re-entering your login details into the wiped device.

Finally, to remove your phone number and any details that associate your device with your network or data carrier, you'll need to deactivate any connected accounts. Contact your carrier for instructions on how to do this.

Tracking a lost or stolen device

If your phone or tablet goes for a walk and doesn't come back, you may think there's little hope of ever having it returned to you, but if you have an app like **WaveSecure** installed you'll have an extra layer of security with the facility to lock down your lost or stolen device, erase the SD card or remove certain permissions.

Lookout can also back up your contacts to their remote servers, although if you already have your contacts synced via Gmail™ you probably won't need to use it. For a yearly $29.99 fee you can upgrade to Lookout's Premium service, allowing you to remotely lock and delete sensitive data from your device; back up and restore photos and call history; and identify any apps which may pose a privacy risk.

Lookout Free ⋅ 4.6

Lookout is a free suite of security tools including a virus scan, firewall and intrusion prevention. It's light on memory use so you shouldn't notice any impact on performance by having it running in the background. But perhaps its best feature is that you can use it to track down your phone if it gets lost or stolen, via an online mapping interface at MyLookout.com. From there you can also set your phone to draw further attention to itself by making a loud siren noise.

WaveSecure can also help locate your phone and then restore much of its contents if recovered. The service costs $20 for a yearly subscription (although the app itself is free), but if you're only after one or two of its features (the ability to locate your phone via GPS, for example) you can often find these for free in other apps, such as **Prey**, or as part of a mobile anti-virus suite such as the aforementioned **Lookout**.

> **Tip:** The best way to keep sensitive data safe is to keep it somewhere other than your phone: on your home computer or backed up to a remote server.

Plan B Free 4.4

If your phone goes missing and you don't have Lookout, WaveSecure or a similar service in place, you may still be able to track it down. Access the Android Market™ website from your computer and remotely install **Plan B** onto the missing device. Plan B is a tracking app from the makers of Lookout that you can install on a device after it's already been lost. Once installed, it'll send details of your phone's location to your Gmail address. You can get updates by texting the word "locate" to your lost phone (from a spare or borrowed phone, obviously).

glossary

Android A–Z

A brief glossary of some of the terms and concepts that crop up throughout this book. Pop back here if we start talking about something you don't understand.

3G, 4G, EDGE

These are different flavours of the same thing: a technology which allows you to access the Internet using mobile telephone networks. It's slower and more expensive than using Wi-Fi but convenient in situations where a Wi-Fi connection isn't available.

Accelerometer

Your phone or tablet most likely has one of these built into it. It's a device that can sense tilt and motion. A lot of apps take advantage of this, allowing you to control them by tilting or moving your handset.

Apk

Apps (see below) are packaged as .apk files. You can think of them as the equivalent of an .exe file in Windows. Some web sources will allow you to download these .apk files and move them to your Android™ device for installation. Others (such as ES File Explorer, see p.108) allow you to back up installed .apk files for future use.

App

Apps are like the computer programs that run on your home computer. They run on your Android device and can perform many of the same functions.

Augmented Reality

Augmented Reality, or AR as it's sometimes known, uses your phone's camera, GPS locator and accelerometer to superimpose a layer of information over the image you see on your screen.

Barometer

Some newer Android devices contain a barometer. Barometers measure altitude, which doesn't seem especially useful for day-to-day use. It does, however, allow the GPS chip to get a much better lock on your location, since it adds an extra dimension (up and down) to the information it's receiving. Barometers also measure air pressure, so who knows? Maybe app developers are already working away at creating home weather forecasting apps.

Bluetooth

A low-range device built into smartphones that lets you wirelessly connect headsets, speakerphones, computers and other network devices (see p.102).

Bookmarks

Bookmarks are shortcuts to web addresses which allow you to revisit a favourite page without having to type the full address back in to your browser bar. Your desktop web browser and the browser on your Android device both use them, and with the right plugins or add-ons you can sync bookmarks from one device to the other (see p.158).

Brick

A phone or tablet that has completely stopped working. Nothing lights up and pushing the buttons has no effect.

Carrier

The company that provides your cell phone or mobile data service.

CPU

The CPU, or central processing unit, is the brain of any computing device. It's the chip where all the calculations take place. Some devices have a separate GPU which controls the on-screen display.

The Cloud

This is just another word for the Internet. When we talk about things being "in the cloud" we're just saying that they exist on the Internet. For example, "cloud storage" is a way of uploading your files to be stored remotely on a web server, instead of storing them locally on your computer, phone or tablet.

Crapware, aka Bloatware

Pre-installed software that a manufacturer or carrier incorporates into your device. Usually this includes games, branded apps and widgets that you'll probably never use and a bunch of perceived "essentials" to get you started. This stuff sits around in your device's internal memory, taking up valuable space. Android 4.0 allows you to disable these apps (see p.116), hiding them and preventing them from loading, but if you want to uninstall them completely to free up RAM, you'll still need to root your device (see p.125).

DRM

Digital Rights Management is a form of copy protection for limiting the use of digital content, such as music and video, and protect it from being pirated or used in ways that the owner of the material does not wish to allow. Some stores and distributors embed DRM into all their wares to prevent them being shared or played using anything other than their own proprietary hardware or software.

eBook

Basically an electronic book format. Amazon's Kindle format is a good example, but eBooks come in a range of other popular formats (see p.198).

Firmware

Your device's firmware is the version of the Android platform that it runs on. See also ROM.

Flash

Adobe Flash adds animation, video and interactivity to web pages. Android 4.0 supports Flash, but future releases won't, as Adobe is phasing out support for it on mobile devices. If your device doesn't already have Adobe's Flash player, you can download it from the Android Market.

Flashing

The process of installing a custom ROM onto a phone or tablet. The actual flashing procedure for doing this varies depending on your device and the ROM you're trying to install.

GPS

A Global Positioning System provides location and time information anywhere on the planet. Your Android device has a GPS antennae built in, so apps can pinpoint your whereabouts on a map, trigger events based on your location, check in to places on social networking sites, and other useful tricks.

HDMI

High-Definition Multimedia Interface, a type of connection that modern TVs, game consoles and audiovisual equipment use. Some Android devices are fitted with an HDMI socket, for hooking up to your TV.

Hotspot

A place that offers wireless Internet access. Hotspots can be found in hotels, coffee shops and various other public establishments.

HTML5

HTML5 is the latest incarnation of HTML – the basic code used to build websites. It allows for multimedia and other forms of content that previously required plugins like Adobe's Flash. Android supports HTML5, giving you a richer Internet experience straight out of the box.

IMAP

An email protocol. Along with POP3, most email clients and webmail services support IMAP.

iOS

The operating system that most of Apple's portable devices run on.

IRC

Internet Relay Chat is real-time Internet text messaging. It allows one-to-one communication via private message (also known as IM, or Instant Messaging) and collective messaging via chatrooms. Popular IRC clients include Trillian, WhatsApp and Imo.

Launcher

The launcher is like a "skinnable" user interface comprising the home screen, widgets and on-screen buttons that you use to interact with your device. Switching launchers is easy (see p.123) and there's a massive selection available from the Android Market.

Live wallpapers

The animated background to your home screen. Your device will have a number of live wallpapers to choose from, or you can choose one from the thousands that are available to download.

Lock Screen

If your device has been left unused for a few minutes the screen will switch off to save power. Then next time you go to use it you'll be presented with a lock screen. Some apps let you install widgets on your lock screen so that you can access frequently used items without having to unlock your phone or tablet (see p.46).

Login

The username (often an email address) and password for any Internet-based services you use.

Malware

Malicious software, an umbrella term which includes viruses, worms, spyware and other evil-doers.

MMS

Multimedia Messaging Service; it's similar to SMS (p.92) text messaging, but can contain photos, formatting and other content.

Multi-Touch

The ability for touch-sensitive screens to respond to more than one point of contact at a time.

NFC, or Near-Field Communication

A chip built into recent Android phones and tablets which allows simple, very short-range data exchange (typically across a centimeter or two) with other devices – another phone, a cash register – anything that can handle an NFC signal. It looks set to become a widely used system for making payments by smartphone. NFC can also be used to read data from museum or retail displays, or to share apps, contacts, photos, music and other media.

Open source

Open source refers to software whose source code is openly available for modification. Open-source software is usually developed collaboratively in the public domain and made freely available.

OS

Operating System. Strictly speaking, Google prefer the term "platform", but for the purposes of explaining what Android does on a phone or tablet, it may be helpful to think of it in terms of an OS, like Windows, Linux or iOS.

OTA

OTA is an acronym for Over The Air: downloading data to your device without having to plug it in. Android system updates are usually OTA.

PDF

Portable Document Format, a type of file invented by Adobe that allows books, magazines and other content to be displayed consistently across a range of devices.

Permissions

When you install an app it will display a list of "permissions". Permissions let you check which of your phone's functions the app will have access to, helping you decide whether or not to allow the app to install (see p.233).

Podcast

A podcast is a downloadable audio or video broadcast recording, usually with a host and/or theme. You can use a podcatcher or newsreader app to subscribe to a feed via RSS. The app will check for new episodes, download them and keep them for when you're ready to watch or listen to them.

QR codes

QR codes are a form of two-dimensional barcode that can contain data such as web URLs or text. The QR codes in this book provide direct links to apps in the Android Market™. If your device has a camera you'll be able to use the **Barcode Scanner** app (available for free in the Market if it didn't come with it pre-installed) to scan these and take you straight to the app, ready for installation. The QR code here is for the aforementioned Barcode Scanner app. The cruel irony is that you'll need the app before you can scan the code so that you can get the app.

RAM

Like a computer, your phone or tablet has built-in RAM where it stores apps and information while you're using them. Unlike a computer, your Android device doesn't have a hard disk, so the RAM doubles as storage space for the operating system and all of your apps (see p.107).

Roaming

If when travelling abroad you want to access the Internet via your phone or tablet's mobile connection service (using 3G/4G, etc) you'll be racking up charges on a data roaming tariff, a separate set of (usually expensive) fees. Thankfully, you can avoid running up high fees by using Android's new data usage monitor (see p.112) to set an upper limit on your mobile data use.

ROM

In the Android realm, ROM refers to the firmware, or operating framework of the platform. Users who "root" their devices (see below) can install a custom ROM such as the popular Cyanogen mod – a modified version of the standard Android release – to provide optimizations and extra functionality.

Root

Rooting is a way of providing "superuser" access to an Android device, allowing you to perform system-level tasks that have, for security purposes, been blocked by the manufacturer. A lot of people root their smartphones in order to install a custom ROM and remove crapware. See p.125 for more details.

RSS

A web protocol that provides updates in the form of a "feed" that users can subscribe to. Newsfeeds and podcasts are often distributed via RSS.

SD card

Effectively a little memory stick that sits in your device, used for storing photos, music, etc.

SIM card

A SIM card is a small removable card that stores your contact details and connection information for your mobile network. SIMs are usually found inside your phone, under the battery. Some modern devices use micro SIMs, a smaller version, and adapters are available.

SMS

Another name for text messaging.

Streaming

A way of playing music or movies across a network as it downloads, rather than downloading a file to your device and then opening it with a program. You can stream media from the Internet or across your own home network from one device to another. YouTube and Spotify are examples of popular streaming services.

Task killer

Task killers, or app killers as they're sometimes known, can be used to force quit an app that may be causing problems with your device.

Tethering

Tethering lets you use your device's wireless or mobile data connection to share Internet access with another device via a USB cable or wireless network. Some mobile carriers and manufacturers disable tethering on their handsets, see p.110.

Trackball

A small sensor that you can use to make fine adjustments when selecting or controlling items on some phones and tablets. Some devices have an actual ball device while others have an optical trackball – the difference being much the same as the difference between a ball mouse and optical mouse.

UI

User Interface: the visual display that you use to interact with your device.

Unlocking

Phones often ship tied to a certain mobile network. If you want to switch to a different network you may need to unlock the device, allowing you to operate it with a SIM card from another carrier. See p.29.

USB

Universal Serial Bus. A standard connection between computing devices and peripherals. The cables come in a variety of sizes but most Android devices use microUSB cables to charge and sync.

USB comes in three versions, the most recent being USB 3.0, which supports transfer speeds of up to 480 mbps, ten times the speed of USB 2.0.

Vanilla

The stock version of Android, without any added manufacturer-specific launcher (such as HTC's Sense or Samsung's Touchwiz) is often referred to as Vanilla.

Webmail

Webmail is the same as email; it's just a means of accessing your email through a web browser rather than through a dedicated email client app, the advantage being you can access it from anywhere. Your email service almost certainly provides a webmail component. Gmail and Hotmail are examples of email services that are web-oriented.

Widget

A widget is a smaller version of an app that can run from your home screen. For example, a weather widget might be periodically downloading and updating weather information to your device; a social networking widget might collect together your contacts' updates from Facebook, Twitter and any other services you're subscribed to and show all of these from an area on the home screen.

Wi-Fi

Wi-Fi is network information broadcast wirelessly from a router. It can be used to share an Internet connection around a home, café, airport, etc.

Index